HONG KONG & TAIWAN
POPULAR DESIGNER SHOWFLATS II
港台当红设计师样板房II

深圳市创扬文化传播有限公司　策划
徐宾宾　三编

江苏人民出版社

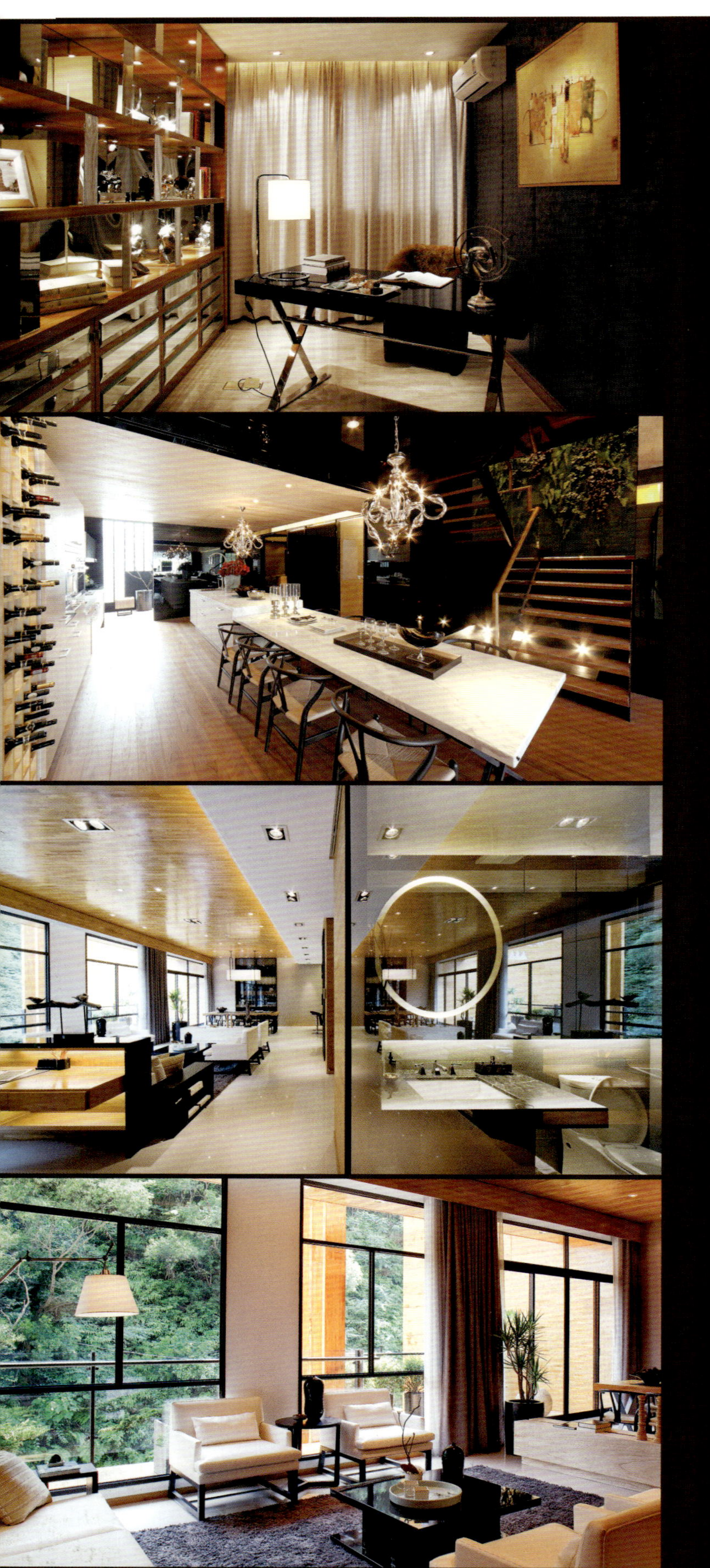

港台当红设计师样板房II
HONG KONG & TAIWAN POPULAR DESIGNER SHOWFLATS II

目录 CONTENTS

Yipinyuan Show Flat
一品苑样品屋

设计师：张清平　设计单位：天坊室内计划有限公司　项目地点：台湾台中　建筑面积：215平方米　主要材料：大理石、镀钛、不锈钢铁件、玻璃、灰镜、木皮、纳米漆

The design takes "less is more" as spirit and sculpture as approach. The reception center deliberately keeps the pureness of a building with light sculpture, simple materials and large areas of pure white. With neat lines, the design tries to present aesthetics of curved surfaces of architecture. Simple materials shape the building into a natural bright luminous body.

The tone is close to modern one, while the style is simple Art Deco, which has formed a transition place from the complication in bustling city. Along with the rhythm of lines, the building has become a modern sculpture full of strengh, showing a futuristic sense in the interlacing light and shadow.

Design focus is not spatial scale or size of the building, but concern over the details. The design has built a harmonious relationship among building, space and people, with extremely exquisite changes of light and shadow, incision, lines and materials.

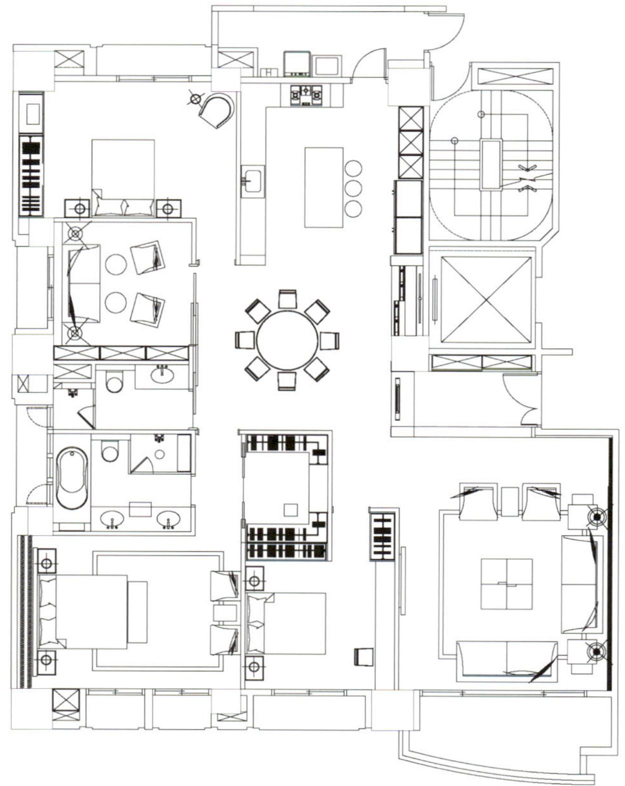

以减法为精神、雕塑为手法塑造，接待中心建筑以轻盈雕塑感的体态、单纯化的月材、大面积的纯白色，刻意维持建筑体的纯净度，以精练的线条展现建筑曲面的美学。简约的用材，将建筑体塑造为一个自然明亮的发光体。

调性趋近现代感，风格则是简约的装置艺术，在喧闹城市里营造出远离烦忧的空间。随着线条的律动，建筑体在光与影的交织间，成为一件充满力量感和未来感的现代雕塑。

设计重点不以空间尺度或规模取胜，而是将重点放在细节的塑造上，以极为细致的光影、切割、线条、建材变化，交融建筑、空间与人的和谐来取胜。

Daan Park Wang's Residence

大安花园王宅

设计师：陆希杰　参与设计：曹车政、郑家皓、刘冠汉　设计单位：CJ STUDIO　项目地点：台湾台北　建筑面积：约188平方米　主要材料：美耐板、面网、橡木集成材、石英砖、盘多磨　摄影师：Marc Gerritsen

There is no excessive decoration on the overall design of this house, which is of unified, pure and clean white tone. It has used the match of light blue veneers and glued laminated wood floor, which has added much visual brightness and created a harmonious tone of wood color and blue and another kind of casual and natural atmosphere. As to the furniture, a set of stone-shaped polygonal modular sofa of boldly used color is placed in the living room, which is free to change. This shows that stunning visual effects can be achieved even in a simple environment. Apart from that, the design has also taken the house owners' life demands and leisure needs into consideration and combined them together. The spatial openness is expressed through different vocabulary. Light blue surfaces have constructed rich sense of depth with lines, which are of free and smooth rhythms.

Study room and dining room is only separated by a semi-open wooden partition. With transparent woven material, the space has maintained good visual and auditory transparent effects. On the other side, frosted glass sliding door and simple and bright passage complement each other, making the space more spacious and comfortable. Then we come to the private bedroom. Perfect proportion of spatial configuration fills the master bedroom with openness. The appropriate storage space, which can play as bedhead, wardrobe, dressing table and partition wall, has added to the utility for the house owners. All of the curved lines meet here and create functional flexibility for the space. The bathroom is built of gray artificial stone. While maintaining original natural flavor, modern and simple features, after treatment, blended with the changes of materials, colors, light and shadow, reveal spatial quality rich in depth. The space can also satisfy users' demands toward utility. The design has combined exquisite details in simple images, and attached importance to life taste besides flexible life changes.

大安花园王宅在整体设计上没有过多的装饰，采用统一、单纯又干净的白色调，运用浅蓝色美耐板与集成材的木地板相互搭配，创造出木头色与蓝色的和谐基调，更营造出另一番轻松自然的新氛围。至于家具设计，则采取用色大胆的石头状多角形组合式沙发摆放于客厅，可自由变化，以此来表现在简单的环境中也能达到惊艳的视觉效果。除此之外，考虑到居住者生活需求和休闲娱乐要求，在空间的开放形式上则运用不同建筑语言。浅蓝色台面以线条勾勒出丰富的层次感，呈现自由又流畅的律动。

书房与餐厅通过一面半敞开式木隔墙联系起来，在视觉与听觉上皆可相互呼应。另一端喷砂玻璃的拉门与简单明亮的廊道产生共鸣，让空间更加宽敞舒适。接着进入私密的卧室空间，完美的空间造型比例，将主卧打造成流线型的开放式空间，再搭配上适度的收纳空间，身兼床头、衣柜、化妆台、分隔空间墙的功能，增添屋主的使用便利性。所有的弧形曲线变化在此交会，为空间延展创造功能弹性。卫浴间则以灰色人造石打造，保留自然原味，现代简约的特性经由处理之后，融合材质、色彩、光影的变化，呈现出多层次的空间质感，并且同时兼顾使用者的生活实用需求。在简单的意象里，结合着细腻的精心安排，除了提供随心所欲的生活变化，也兼顾品位生活。

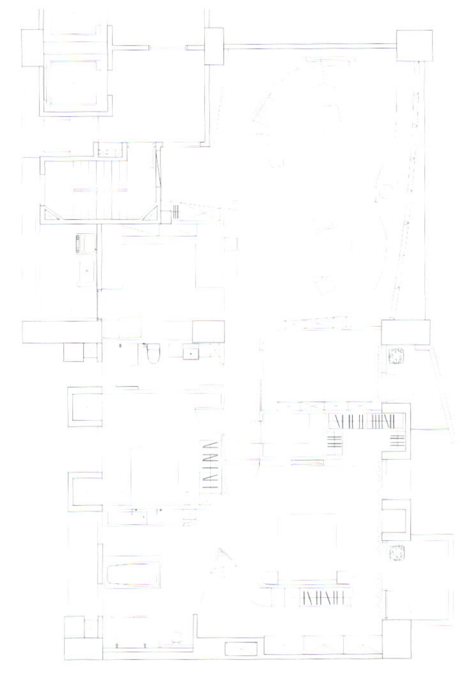

Zhubeiyuan Jianzhu Show Flat

竹北原见筑样品屋

设计师：陆希杰　参与设计：刘冠汉、张琼之、林丽纹　设计单位：CJ STUDIO　项目地点：台湾新竹　建筑面积：约172平方米　主要材料：盘多磨、南方松、卡拉拉白、马赛克、钢琴烤漆、乱纹不锈钢　摄影师：Marc Gerritsen

Designers have applied approaches like division and partition to create this space. Entering the space, you will first be greeted by an entire bright green wall. Although there is not any other decoration on the wall, except a small wall clock, bright color of the wall has greatly lit up the whole space and enriched visual space. The overall space takes white, black and gray as basic tones, together with wood color furniture, green potted plants and furnishings, making the entire space lively.

An open kitchen allows a more harmonious relationship between people. The clean white kitchen stimulates people' s desire to have a try in cooking. The most special place in the whole space should be the leather sofa that is set along the whole wall, whose flowing curves make the space look smarter. The partition between the bedroom and bathroom is only a white wall, which not only separates the two rooms, but also connects them. Black, white and gray bedding is complementary to the black and white painting at the corner. The overall atmosphere in the space is sedate and leisurely. In the bathroom, the house owner can wash away the day' s fatigue and read some pages under the bedside lamp and then slowly fall into his sound sleep.

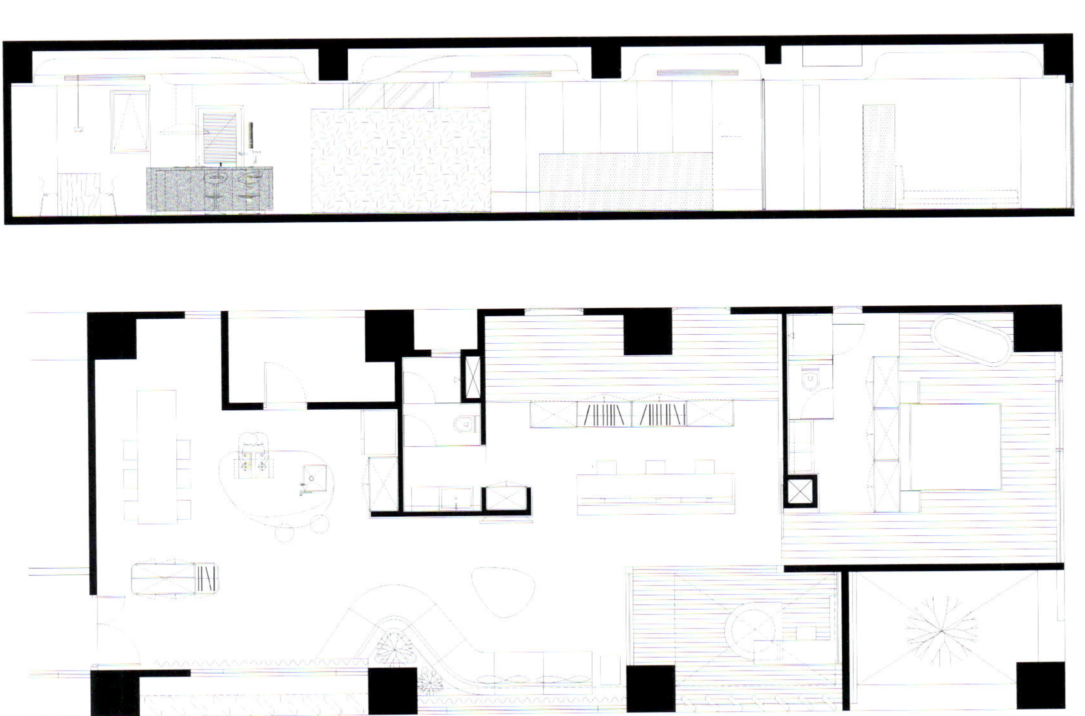

设计师以切割、隔断的手法打造此空间。步入室内，首先映入眼帘的是那一整面亮绿的墙面，虽然墙面没有其他装饰，只是挂着一个小小的钟，但墙面那明亮的颜色就让整个空间增色不少，也丰富了视觉体验。整个空间以白、黑、灰为基调，再配以原木色的家具和绿色盆景、饰品，让整个空间都活跃起来。

开放式的厨房，让人与人之间的关系更为融洽。白色干净的厨房，让人有跃跃欲试的欲望。整个空间最具特色的地方要属那沿着整面墙而设的皮质沙发，流动的曲线使空间更为灵动。卧室与浴室也仅一面白墙之隔，却隔而不断。黑、白、灰色的床上用品与墙角的黑白画作相呼应，整个空间表现出一种沉稳、慵懒的气氛，主人在浴室洗去一天的疲惫，在床头灯下翻看几页书，慢慢睡去。

The Show Flat of Haiyiwanpan Building 2 6CD

海怡湾畔2栋6CD样板房

设计师：Thomas 、Wing　设计单位：戴维斯(国际)设计及顾问有限公司　项目地点：广东深圳　建筑面积：150.7平方米　主要材料：云石、天花、玻璃、镜子和不锈钢

This case is of modern minimalist style. The design is not confined to lines and frames. The clear visual effects bring the room the unparalleled feelings, and the neutral color of beige and brown is scattered in the room. The marble is utilised in the living room, the dining room and the study, while the natural teak occupies other spaces.

The tough and hard interior design brings out the unexpected feminine elements. Each room has its own decorative features, and decorative elements are used in the ceiling, the walls, and the furniture. The glass, the mirror and the stainless steel bring out the fresh and fashionable style, which is the perfect blend.

此案例为现代简约风格。设计不限于条框，清晰的视觉效果给人一种无可比拟的强烈感受。采用中性的米黄色与棕色。云石在客厅、饭厅与书房中随处可见，天然柚木则占据了其他空间。

阳刚硬朗的室内设计中漫溢出柔软的特质。每个房间都有各自的装饰特色，这些装饰元素被运用在天花、墙面、家具上，在玻璃、镜子和不锈钢的衬托下佤现出清新时尚的风格，并完美地结合起来。

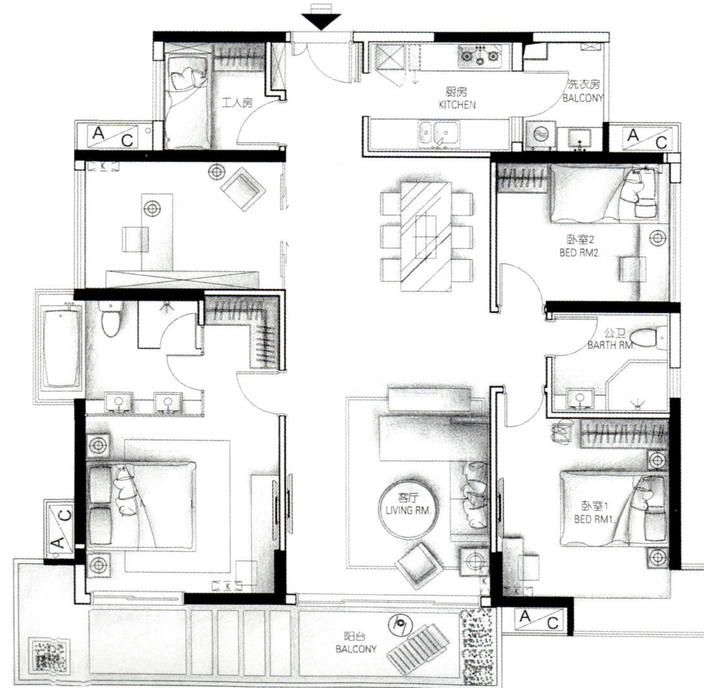
工人房
厨房
KITCHEN
洗衣房
BALCONY
卧室2
BED RM2
公卫
BARTH RM
客厅
LIVING RM.
卧室1
BED RM1
阳台
BALCONY

The Show Flat of Haiyiwanpan Building 2 6E

海怡湾畔2栋6E样板房

设计师：Thomas 、Wing　设计单位：戴维斯(国际)设计及顾问有限公司　项目地点：广东深圳　建筑面积：98平方米　主要材料：大理石、木材、玻璃、皮革

This case is an elegant combination of dignity and popularity. It attaches great importance to optimizing the viewing platform at high floor where you can enjoy the sea views. While the interior design focuses on the enhancement of the room's comfort so as to make it an ideal high-end residence for an elite group who pursues a quiet life away from the hustle and bustle of the city.

In order to maximize this advantage, the layout of the indoor furnishings has been carefully selected to create a quiet, luxurious, and modern atmosphere. The extensive use of soft beige and black symbolizes a living attitude of Armani-style.

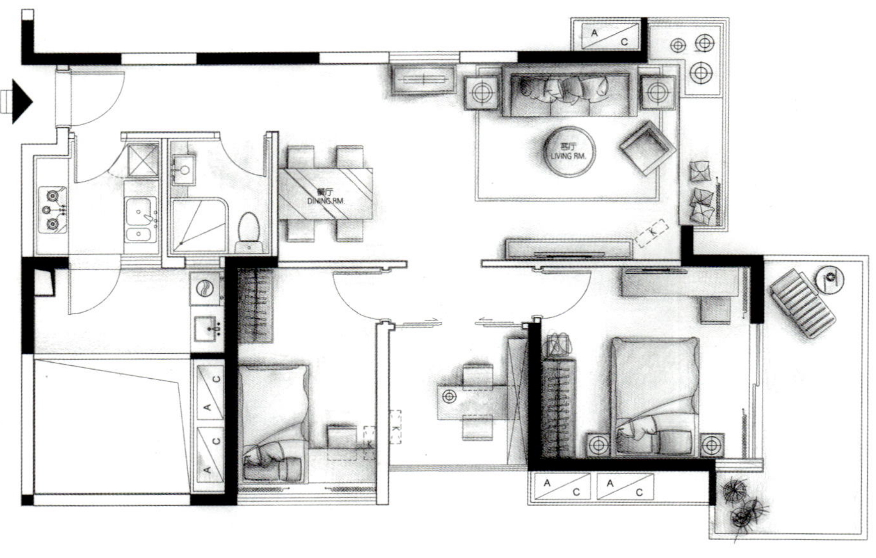

此案例为尊贵与流行相结合的雅致典范。设计重点是优化这座位于高层的观景平台，在这里可以欣赏富有诗意的海景，内部设计重在提升房间的舒适度，使其成为崇尚宁静生活、远离城市喧嚣的上流人士的理想高端住宅。

为了使这一优势得到最大限度的开发，屋内的家具摆设都经过专业人士的精挑细选以营造一种宁静、奢华、现代的氛围。柔和的米黄色与黑色的结合运用，象征着一种阿玛尼式的生活态度。

Parkland No.26 Building

公园大地26栋

设计单位：戴维斯(国际)设计及顾问有限公司　项目地点：广东深圳　建筑面积：202平方米　主要材料：玻璃、大理石、软包、茶镜

The delicate and elegant living space has become the ideal home for those urban new aristocrats. A chic and modern space is successfully created by designers who pursuit delicate details and well organize the materials such as glass, tawny glass and marble in this case. The whole space exudes a charming flavor of vogue and sumptuousness.

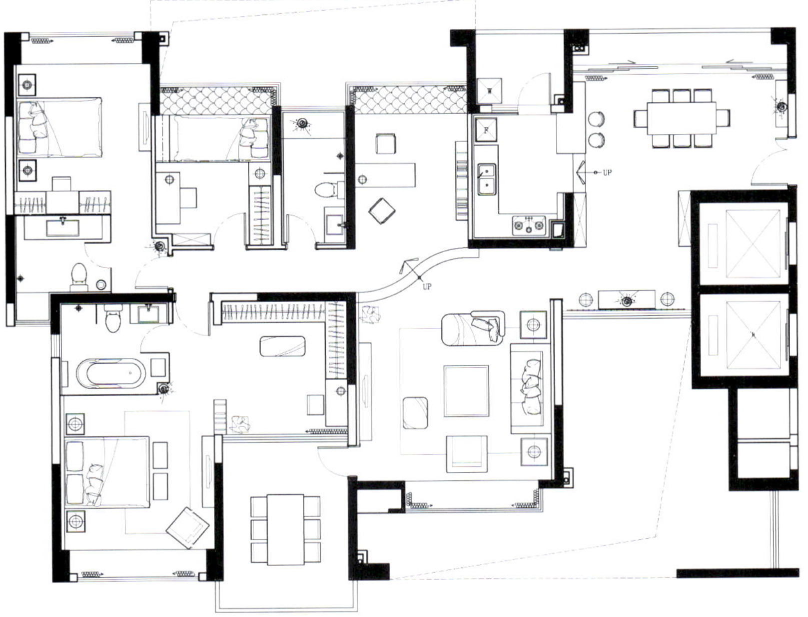

精致典雅的生活空间是时下都市新贵所追求的理想之家。此案例设计师通过对精致细节的追求和对玻璃、茶镜、大理石等材料的合理运用，成功地打造出一个雅致时尚的生活空间。整个空间散发着一种迷人的摩登奢华的韵味。

Parkland No.32 Building

公园大地32栋

设计单位：戴维斯(国际)设计及顾问有限公司　项目地点：广东深圳　建筑面积：200平方米　主要材料：石材、马赛克、大理石、玻璃、茶镜

The design of this show flat is not confined to the rules and regulations, and is modern and classic. The clear visual effect brings out the incomparably strong feelings. Slim and graceful colors show immeasurable glamour. Rough or smooth, matte or glaze, dark or light, by using all these materials and colors, designer shows a balance in detail, creating a harmonious interior space.

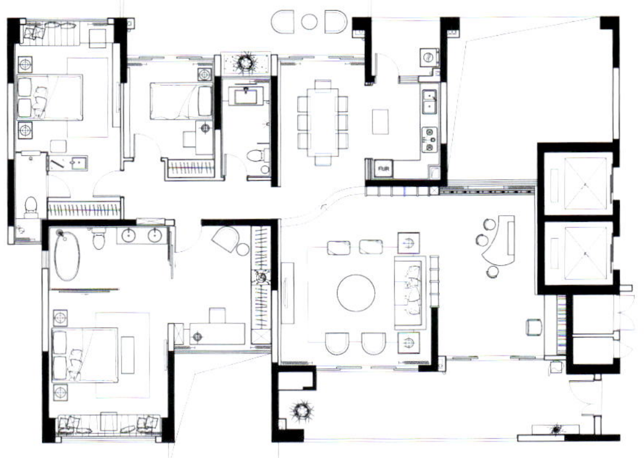

现代、华贵而经典，设计不限于条框，清晰的视觉效果给人一种无可比拟的强烈感受。轻盈飘逸的缤纷色彩，散发出无限的魅力。粗犷与平滑，哑面与光面，深沉与明亮……各种物料、色彩相互融合，展现出细腻的平衡，以营造和谐的室内空间。

Parkland No.36 Building

公园大地36栋

设计单位：戴维斯(国际)设计及顾问有限公司　项目地点：广东深圳　建筑面积：123平方米　主要材料：玻璃、茶镜、水晶珠帘、地毯

It is characterized by the unique decorations and fragrance. The hard edging and smooth surface, as well as coffee and exquisite metal crystals, create a charming and modern house. It is a commensal unit with an open living room and multi-functional rooms, with drawing room, dining room, kitchen, bar and study room integrating into one.

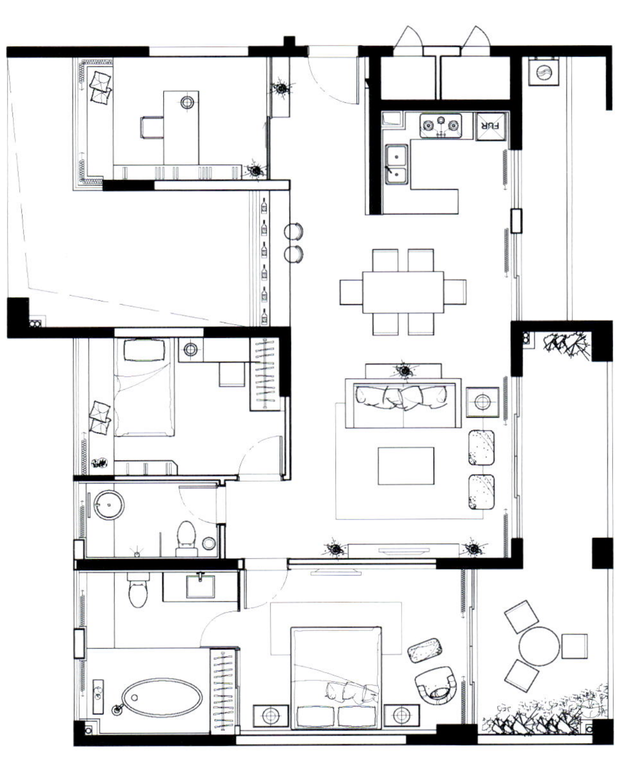

以创新独特的琳琅馥郁为特色，加上硬朗的边饰与光滑的表面，再结合深咖啡色和精致的金属水晶，塑造出动人的迷醉空间。整个空间设计构建出一个共栖式的生活单元，它拥有开放式起居及多功能空间，客厅、餐厅、厨房、酒吧、书房合成一体。

DIET CULTURE

Always with joy

Show Flat at World Trade Center in Wuxi

无锡世贸中心样板房

设计师：梁景华 设计单位：P A L 设计事务所有限公司 项目地点：江苏无锡 建筑面积：约130平方米

The one of the three sample houses located in the CBD district in Wuxi city, emphasizing the spirit of traditional Chinese culture, shows its comfortable and fashionable decorations in a contemporary and modern approach to the potential buyers of the mansions. It integrates the traditional Chinese craftsmanship, and uses the modern grey as the main color. It is simple and delicate against the Chinese elements such as the translucent screen with lucky patterns.

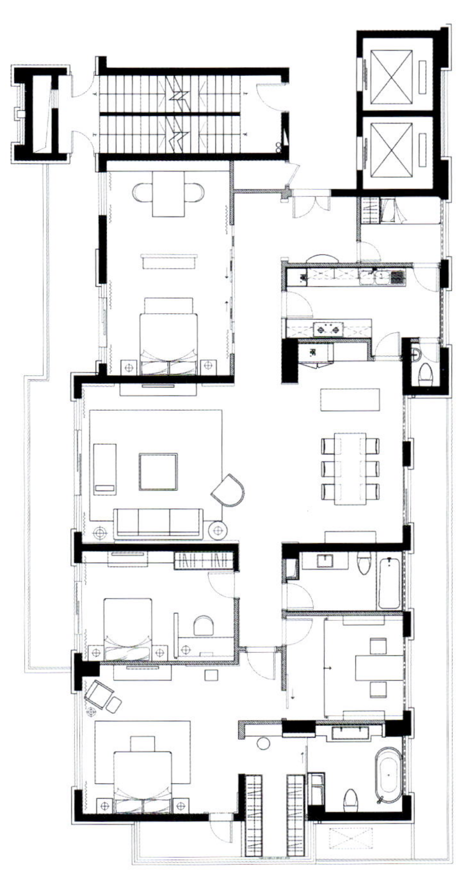

这是无锡世贸中心的三个特色样板房其中之一，注重中国传统的精髓，以现代中式的设计表现舒适的时尚家居，呈现给豪宅的准买家，糅合东方哲学的二艺雕琢，颜色却以现代感的灰色调为主线，配合中式元素的吉祥图案透光屏风，简约精致之感随风散发。

3D Light Rail—Dazzling Futuristic Space

3D光轨——炫目未来感空间

设计师：马健凯　设计单位：界阳&大司室内设计　项目地点：台湾中和　建筑面积：约96平方米　主要材料：毛丝面不锈钢、镜面不锈钢、亚克力、黑色烤漆、明镜、灰镜、银狐大理石、石材、灰镜马赛克、水晶灯、文化石

The design style is avant-garde and fantastic. What catches eyes first is the extremely dazzling "light" which is inspired by the movie TRON: Legacy. The space has entered a virtual world of science fiction movie, where laser light rails interlace with each other and form tension of picture. Beams of light penetrate through plates made of special material. Mirror stainless steel has framed the screen partition. Front and back layers of light have composed three dimensional futuristic images, bringing new visual impact.

The magnificent wall that leads perfectly straight from the entrance hall to the living room is designed with more than eight kinds of fashionable materials. Harmonious collocation approaches highlight the "disguise" concept which JIE-YANG Interior Design is good at. Consistent stylish design has covered and hidden the necessary storage function. As a show flat of JIE-YANG Interior Design, the backdrop of living room has a horizontal structure. Painted glass engraved with "JIE-YANG INTERIOR DESIGN" is presented in a low profile on a stainless steel plate, which has become a brand logo of the show flat.

The space is of fashionable lounge bar style. The main wall made of culture stone in the dining room is horizontally embedded with mirror stainless steel, which has increased the material variability. Below the luxurious crystal chandelier is a custom-made frosted-surface stainless steel dining table. Glass material plus LED light creates a relaxing atmosphere. What is special about the design is that designer Ma Jiankai has intentionally chosen the single chair + bench combination, which is out of careful consideration for the neighboring entrance hall. A bench without back can reduce the repression brought by solid form and can also function as a shoes bench.

Some walls are dismantled and the configuration has been changed into 2+1 rooms. Transparent glass house has stretched and enlarged the space visually. There are rows of wardrobes hidden, and bedplate can also be opened, which has endowed the space with more flexibility. To present more design styles, JIE-YANG Interior Design has designed the two bedrooms of completely different styles. The master bedroom is a cold tone, black space, in which a gray mirror mosaic wall can reflect colorful light. Stylish and luxury vocabulary has been infused into the space to highlight the dazzling main body with a complete black room. In the guest room, the extreme design concept has turned into white fashion to set off the black wall with completely pure white. To fit for the style, a stainless steel tree-shaped clothes rod is placed in the corner, which is consistent with the overall fashionable taste.

CHANEL

CHANEL

前卫虚幻的设计风格，首先攫住视线的是炫目亘极的“光”，来自电影《创：光速战记》的灵感，进入科幻片的虚拟世界，激光光轨交错在眼前的画面张力，光束穿过特殊板材，再以镜面不锈钢框住屏风，前后光线层次构成立体的未来式印象，带来新的视觉震撼。

对于由玄关笔直往客厅延伸的大气墙面，设计师则利用超过八种潮流素材，协调而不显零碎的搭配手法，强调界阳室内设计所擅长的“伪装”概念，用一贯的时尚设计感，包装、隐藏应有的收纳功能。既然是界阳设计的展示样品屋，客厅背墙以虚实衔接的水平架构，在不锈钢板上低调展示刻有“JIE-YANG INTERIOR DESIGN”字样的烤漆玻璃，成为样品屋的品牌标志。

时尚休闲酒吧的风格主题，餐厅文化石主墙横向嵌入镜面不锈钢，增加材质变化。在奢华的水晶灯下方，定制的毛丝面不锈钢餐桌采月玻璃面材结并合LED灯光，营造适合举杯小酌的放松氛围。特别值得一提的是，设计师刻意使用单椅+长椅的组合，因为邻近玄关入口，无背式的长椅既能减少对形体的压迫，还兼具穿鞋椅的功能。

拆除实体墙改为2+1房，清透的玻璃房延伸、扩大空间，不但藏有成排的衣橱，还使床兼具收纳功能，赋予空间更多元使用的弹性。界阳室内设计为了呈现更广的设计面，两间卧室以截然不同的风格呈现。主卧室为黑色，一面折射七彩光芒的灰镜马赛克主墙，融入时尚奢华语汇，以全室素黑来映衬绚丽耀眼的主体。极端的设计概念至客房，转为白色时尚，以天然的纯白色来烘托黑色墙面，角落妙搭不锈钢材质的树状吊衣杆，延续一贯的时尚品位。

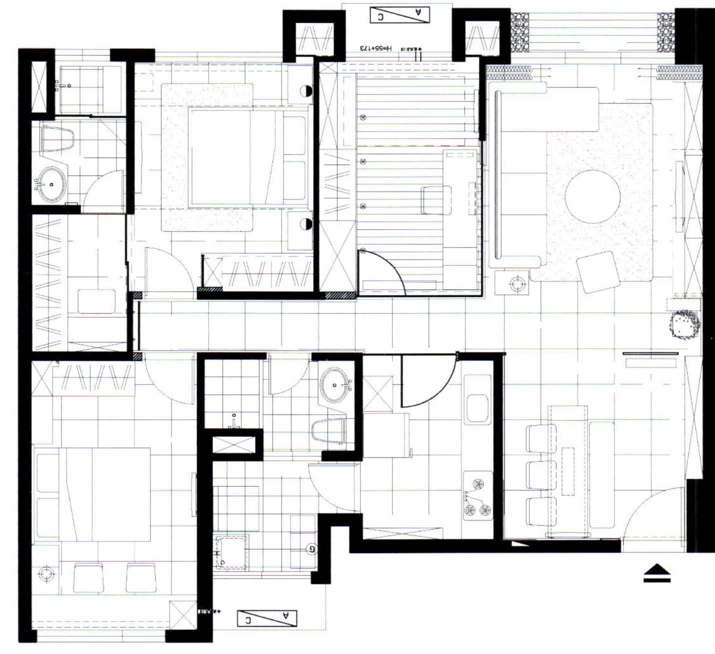

Roaming in the Glittering Space of Light Properties

漫游潋滟光感空间

设计师：马健凯　设计单位：界阳&大司室内设计　项目地点：台湾桃园　建筑面积：约175平方米　主要材料：镜面不锈钢、钢琴烤漆、进口壁纸、亚克力、LED

Avoiding adopting the bright surface of the stainless steel screen, this case emphasizes the trendy and avant-garde originality and process so that the space glitters sparkling radiance just as water waves. JIE-YANG Interior Design makes the living space for young people stylish, decorated with shinning surface and woolen stainless steel surface to refract swimming kinetonema of light and shadow. The kinetonema sometimes bursts into cold lights going through the white view wall, and sometimes turns into 3D laser light rails which stagger in the bar. Each field is exotic, such as the dance floor of the nightclub, the LED colorful floor in the locker room. This commercial space is designed to allow an open locker room to change into a fashion runway for personal show.

The curved TV wall with three levels extends toward the center, gradually narrows and at different angles. The upward linking ceiling lines easily bring out the habitants' roaming paces. The creative linear dynamic extends to the formed desk in the study, starting with the tea table in the living room, which defines the boundary between the living room and the study through the upward, twisted, extended and smooth decoration at one stretch. For light atmosphere between the two spaces, the design director adopts especially custom-made lighting equipment to reflect the beams of light scattered in the central facade, which becomes a unique decorative symbol.

In the boutique master bedroom with youthful flavor, the white light box at one side of the entrance displays the couple's collections. And the surface of the semi-high video wall is divided into triangular blocks which are painted with black and white to strengthen the visual contrast effect and decorated with stainless steel frames to add the texture. The bold design for commercial space is applied to the locker room with the nightclub style for the young couple. The colorful light from LED hidden under the ground brings us stylish atmosphere from time to time.

Irregular curved layers continue its theme in the furniture. The creative torsion of stainless steel blooms out beautiful waves in the curved surface, which soften the masculine texture of the materials. JIE-YANG Interior Design makes some innovation based on the design elements loved by young people to create an avant-garde fashion in black and white. Featuring a soft and relaxed feeling, this is a residential space with a unique specificity.

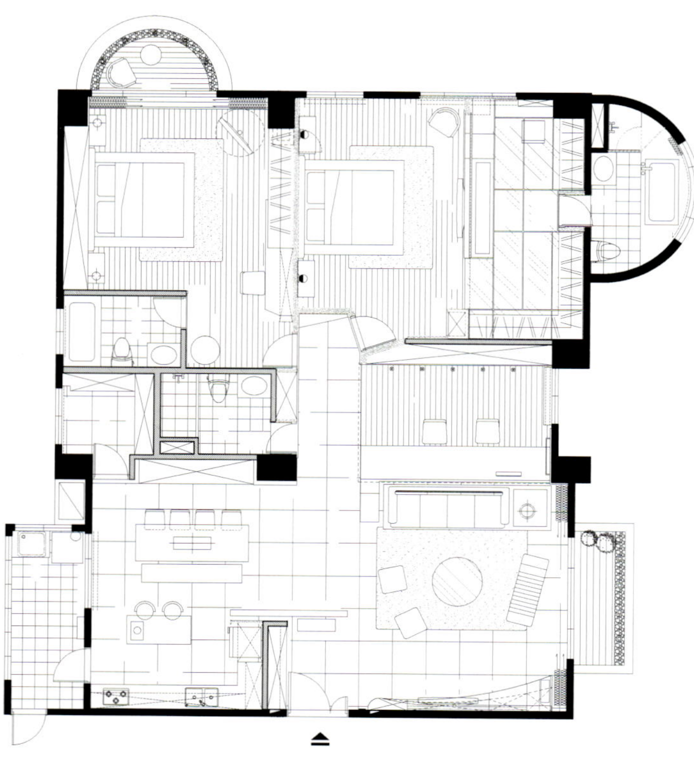

扭转的亮面不锈钢屏风，闪耀着水波潋滟般的光彩，创意新潮且前卫，界阳室内设计在年轻人的居住空间大玩时尚，多处点缀亮面以及毛丝面不锈钢，折射光影的嬉游动线，时而绽放冷光穿梭于白色端景墙，又幻化为3D激光光轨交错于吧台，在各个区域皆是领衔之姿。最终，如有炫光效果的夜店舞池、从容踩踏于更衣室的LED七彩变幻地板，商业空间的设计强烈度，让开放式更衣间成为了一个人秀的时尚伸展台。

圆弧状的有三个层次的电视墙，其深度向中央渐窄，以不同的角度延展，向上连接天花线条，轻松带出居住者的漫游脚步。创意的线性动态，延续至书房一体成形的书桌，以客厅茶几为起点，上扬、转折、延展，流畅地定义出客厅、书房的分界线。而在两个空间的灯光氛围，设计总监定制特殊的灯光设备，折射冷色光束散落于中央立面，成为独特的装饰符号。

年轻化的精品主卧房，入口一旁的乳白灯箱，高调展示夫妻俩的收藏品，半高影音墙的表面分割成三角块状，黑白填色对比出视觉强度，并框饰不锈钢，平添质感。用于商业空间的大胆设计，套用到年轻夫妻的夜店风更衣室，地面藏入LED七彩光影，不时变换时尚氛围。

不规则的弧形层次，在家具上也可找到；在赏玩不锈钢的创意扭转时，在弯曲面绽放美丽波光，弱化了材质的阳刚味。对于年轻人喜爱的设计元素，界阳室内设计对其加以变化创新，黑白前卫的时尚风潮，以柔软轻松的感受调味，创造住宅空间独一无二的特性。

Small and Special Space

方寸之间

设计师：郭宗翰　设计单位：M. Design石坊空间设计研究事务所　项目地点：台湾台北　建筑面积：150平方米　主要材料：石材地坪、木地板、石材、实木木皮、玻璃、铁件　摄影师：Kyle Yu

This project takes gray as the main tone, providing people with kind of sedate and leisure feelings. Design of the entertainment area is concise and comfortable, without too many decorations. Log wall, white flowers and green plants give off some natural aura. Four rectangular forms are engraved on the log wall in the passage way with hollow approach and are decorated with glass, visually enlarging the space and making it more transparent. A small bookcase is placed between two walls. Looking from the distance, you will feel something special is hidden in the small bookcase. Light in the dining room is extremely modern, whose flowing curves make the space especially smart. Door of the kitchen is a sliding glass door, which has enhanced the visual depth. The whole space is presented with approaches that can make it more transparent. The lines are neat and clear, which has showed designers' infinite creative ideas in this small but special space.

此案例设计以灰色为主调，给人一种沉稳、慵懒的感觉。休闲区的设计简洁、舒适，没有太多的笔墨，点到即止。原木的墙面、白色的花朵、绿色植物，透露出些许自然的气息。走道的原木墙面上以镂空的手法雕刻出四块长方形体，以玻璃装饰，使视觉得到延伸，空间更为通透。两面墙之间的小型书架，从远处望去，令人觉得内有乾坤。餐厅的灯具，极富现代感，流动的曲线使空间灵动无限。厨房门采用滑动玻璃门，加深了视觉的延展。整个空间都以通透的手法展示在人们眼前，线条干净利落，在这方寸之间，展现了设计师的无限创意。

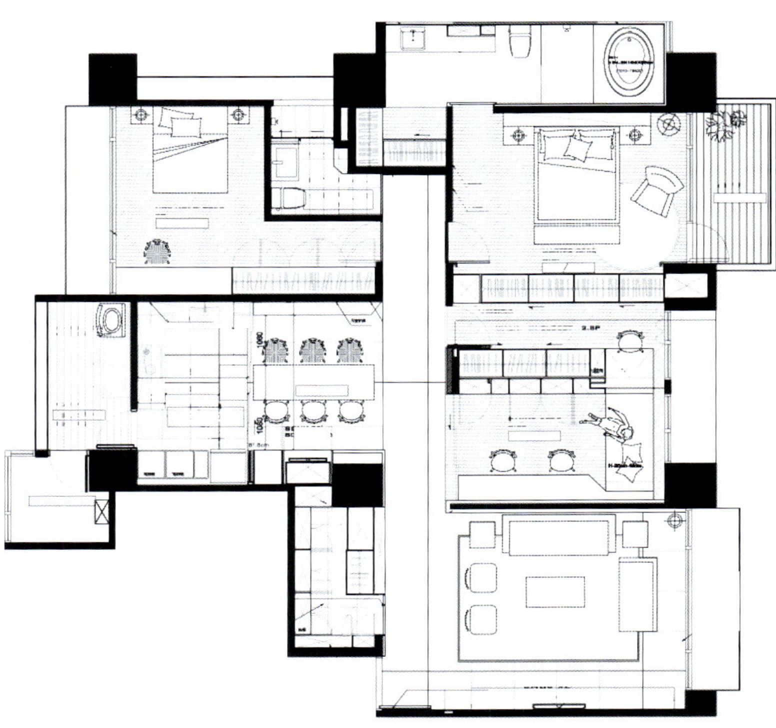

Cai's House, Banqiao

板桥蔡公馆

设计师：虞国纶　设计单位：格纶设计　项目地点：台湾新北　建筑面积：100平方米　主要材料：琥珀镜、明镜、咖啡色烤漆玻璃、银狐石、乳胶漆、仿古金漆、白色钢琴烤漆、进口珠光壁纸、贴金银箔定制家具、缎面绷布

Fashion and neoclassicism are defined just like cutting diamond. The lines are brought out by bright mirror at the entrance. And these lines run from the wall to the ceiling, magnifying the space and strengthening the frames of different areas. The folding screen at the entrance is framed with bevel angle of bright mirror, which is a precise match and even catches up with the turning result of diamond. Matching with the baking finished glass in the center of screen can create a different magnificent atmosphere by this low-key difference. This time, the designer makes the white paint as the base matching with bright mirror, grey mirror, amber mirror, baking finished glasses, letting the whole space present the diamond like shine by the delicate cutting in different mirrors, smart light arrangement and different materials.

The designer creates the classic case once more by the outstanding match. Natural pattern of silver stones show natural appearance in the TV background wall in the living room, and reflect delicate and grand looks under the shadow change of day and night.

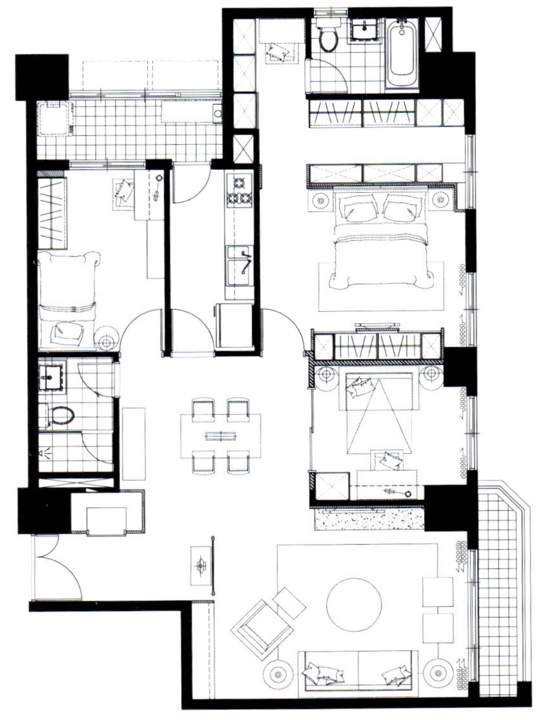

以钻石般完美切割，界定时尚与新古典。玄关处以明镜拉出线条，由壁面转至天花板，将空间放大，加强了不同区域之间的界定。玄关入口处的屏风，以明镜倒斜角作为框边，精准的契合，谁说只有钻石的切工才够完美！搭配上屏风中央的烤漆玻璃，这种低调的反差，营造出与众不同的华丽氛围。设计中以白色喷漆为基底搭配上明镜、灰镜、琥珀镜、烤漆玻璃，让整个空间在不同层次的白之中以不同性质的镜面细腻的分割、巧妙的光线布局，在不同材质的巧妙搭配下闪耀出犹如钻石般的璀璨光芒。

设计师以细腻且不流于俗套的搭配再一次成就了经典。以银狐石材的自然纹路，在客厅的电视主墙显现出自然的动象，在日夜、光影的更迭下，呈现出精致与大气的各种面貌。

Scenic Show Flat

观景样板房

设计师：陈鹏旭 王文炜　设计单位：光合空间设计　项目地点：台湾

Nature is the place where the heart returns. It' s a true design whose spatial arrangement focuses on human and natural life. This case is themed with nature' s existence in living room, paying attention to the union of human, environment and architecture, which meets the requirement of people about natural life. The mood can be nice with wind, light, water and green, which can realize the dream to live an ideal life in city.

This case is facing Yuanshan Mountain, Keelung River, Huabo Park from the far to near, and introduces scenery from nature, making harmonious union of hills, water, trees and buildings, extending the scenery and making it endless. Balcony and loft stairs feature vertical 3D plants wall and shallow pool to communicate with trees outside. The private garden even lets the space lively. Walking inside the green, people can smell the fragrance from the flowers, and then the outdoor can complement the indoor.

The content of plane can be illustrated from the aspect of life style. The continuous moving lines downstairs shape the space full of flowing inspiration. There' s a corridor free design upstairs. From the open space to the respectively private bedroom, the elastic living room in the central section, promotes the perfect interaction between individual and family.

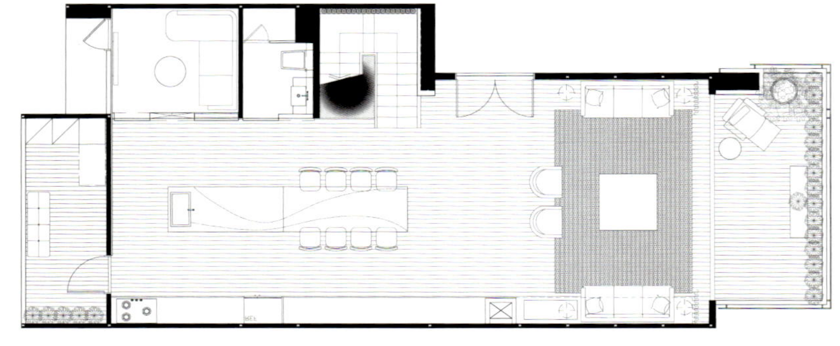

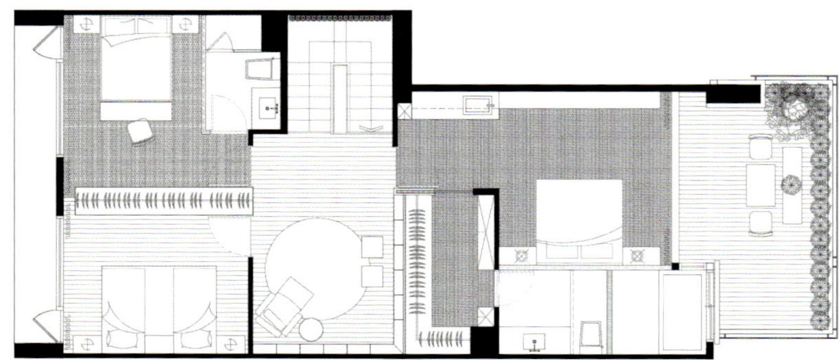

自然是心灵的归属，以人与自然生活为中心思维的空间规划，才是真设计，以自然就在家中厅堂为主题，考虑人、环境、建筑合一，满足人们对于自然生活的向往，让心境随着风、光、水、绿一起畅游，品味在都会耕植理想生活的隽永风景。

此案例基地由远而近面对圆山山脉、基隆河、花博公园，以三度空间思维，向大自然借景，让山、水、树、建筑和谐融合，景色无限延伸。阳台、跃层梯间以垂直角度的立体植生墙、清水浅池设计和户外浓荫对话，私属花园更增添了空间的生命力，让行走间融入绿意，延伸绿树视觉与花香嗅觉，于是室内与户外有了最美的呼应。

从生活形态规划平面的内涵，楼下连续的动线塑造整体流畅的空间。楼上无走道式设计，实现开放空间到独立私密卧房的转换，正中央创造弹性的枢纽起居室，提升个人与家庭的互动。

Classic Photosynthetic

经典光合

设计师：陈鹏旭、王文炜　设计单位：光合空间设计　项目地点：台湾台北　建筑面积：264平方米　主要材料：集层木皮板、黑网石石材、灰玻璃、强化玻璃、金属建材

All designs come from the extension of the conception about peaceful coexistence with nature. In the design of "Jiangshan", we've seen more completed natural narration from TWSDI. The design stars from the scenery. The base originally is the natural protection area facing the inner lake, and surrounded by hills. Therefore, the designer makes the original nature be the leading role and pulls the building inward, creating a beautiful picture brought by present hills and the building. Besides, the designer also decorates the entrance wall with the irregularly cut stones which are strewn at random. When sunshine is on them, the shadow will scatter on the ground, just like that created by woods where people walk inside.

In the public area, the designer reduces the decorative elements to the minimun by silent transmission with the precondition without disturbing or destroying the harmony. Get rid of the complicated ornaments and return to the purest heart. The top floor is themed with fantastic scenery of the hills. The wooden stairs are lifted and big trees are planted beside to complement the opposite green mountain.

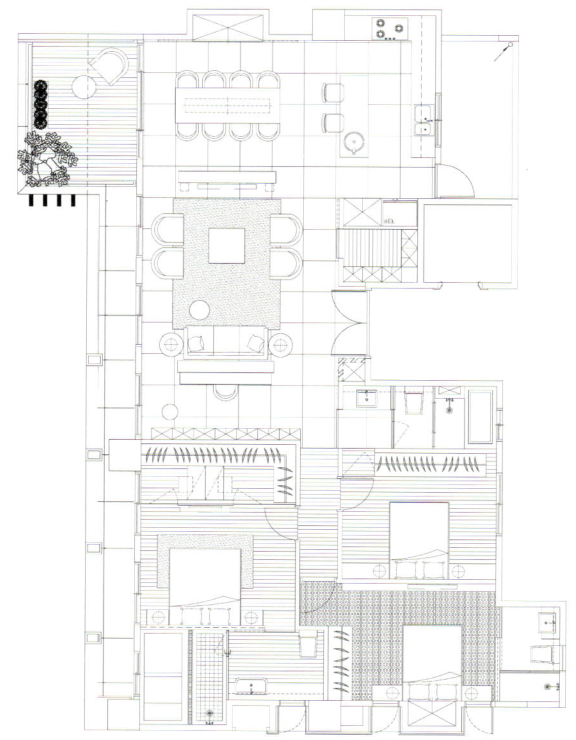

一切设计皆源自于与自然和平共存的概念，在“疆山”之中，我们看到光合空间设计更完整的自然叙述，从景观开始，由于基地本身即面对内湖的自然保护区，四处皆为群山环绕。因此，设计师让原本的自然生态成为主角，将建筑物本身往内退缩，从而借既有的山景和建筑整体营造出水之美的意象，并在围墙上花费心思，运用石材的不规则切割，将之错落地摆设，形成大小不等的交织，当阳光照耀其中，光影洒落地面，便如同是树林营造出的阴影一般，仿佛游走在树林之中。

在公共区域之中，设计师借由宁静的无声传递，将一切人为的装饰性元素降至最低，以不干扰、不破坏的维持和谐为前提，拿掉华丽的赘饰雕砌，回归到心灵最干净的原始状态。位于基地的顶楼，以山林美景为主题，将木阶抬高，在一旁种植大树，呼应对面的山景绿意。

Fashionable Fun

时尚嬉趣

设计师：房元凯　设计单位：凯奕设计顾问有限公司　项目地点：台湾台北　建筑面积：125平方米　主要材料：彩色砂岩、金属烤漆、烤漆玻璃、进口特殊瓷砖

The charming fashionable elements and modern image inside are the microcosm of city nightlife. The designer frames the colorful city and interprets the life in an artistic way. The design represents the special aesthetic idea and it is the feature of the space.

The designer is good at the use of light to form a dreamy stage and to express feelings in an invisible way, stirring up fashionable nerves inside. It changes the light, projection light, fill-in light and exquisite light source into the index. The house is shining and modest against the contrast of light and shadow.

此案例拥有令人着迷的时尚元素，室内的摩登意象，是都市夜生活的缩影，设计师框住都会绮丽境地，以当代美学架构层次，将生活风格延伸至艺术气息，设计手法诠释独特美学主张，凝聚成空间特色。

华灯初上，光影迷离，构成舞台场景，这是无形的情感表达，挑起隐含于内的时尚气息，将打光、投射、接光、补光、细腻的光源元素，扮演区域变化的索引，明与暗、衬托与反差，随着不同方向折射的光影，各式置入的物体时暗时明。

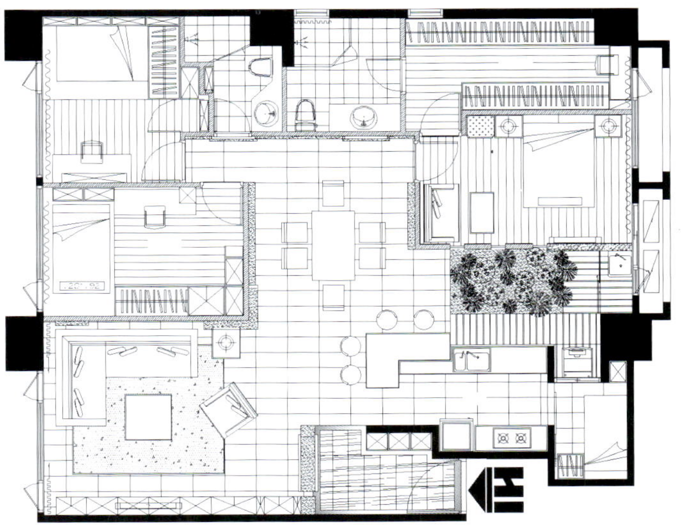

Yale T5 Show Flat of Yuanxiong University
远雄大学耶鲁T5样品屋

设计师：黄书恒　参与设计：欧阳毅、陈佳琪、王建益、胡春惠、胡春梅　设计单位：玄武设计（Sherwood Design）　项目地点：台湾台北　建筑面积：约100平方米　主要材料：卡拉拉白、橡木皮、黑色烤漆玻璃、特殊壁纸、黑色橡木地板、地毯

Simple grey shows extraordinary glamour. Indigenous flavor and fashionable patterns are found everythere. The dining room integrated into living room can be arranged well. This kind of dining room mainly focuses on the function and aesthetics. At the same time, the main space here is living room whose decorative style must be correspondent with that of living room. Otherwise, these two connected space will bring a disconnected feeling.

The match of black and white gives us a strong visual impact. Big areas of black paint, black TV, and the united stereo system contrast with light color, giving off a fashionable flavor. Beige sofa going well with simple pattern dotted back-pad make people feel comfortable and warm, which are simple and cozy. The main key of white has a strong reflecting ability, making the room brighter.

With fashion design, outstanding and eye-catching, proud or low-key, all are pursuing distinguished personality; the pure basin and toilet make this small wash room bright and transparent.

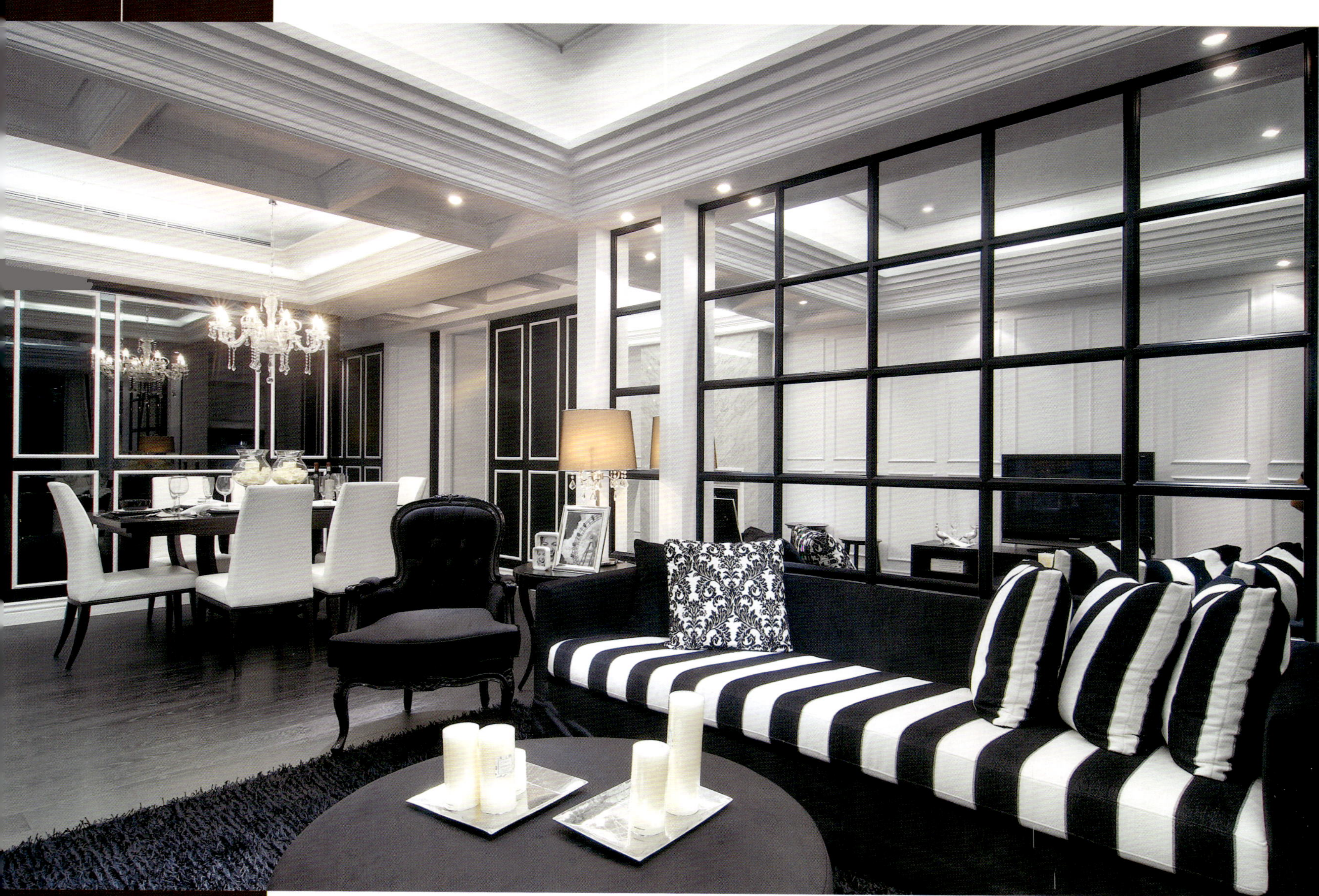

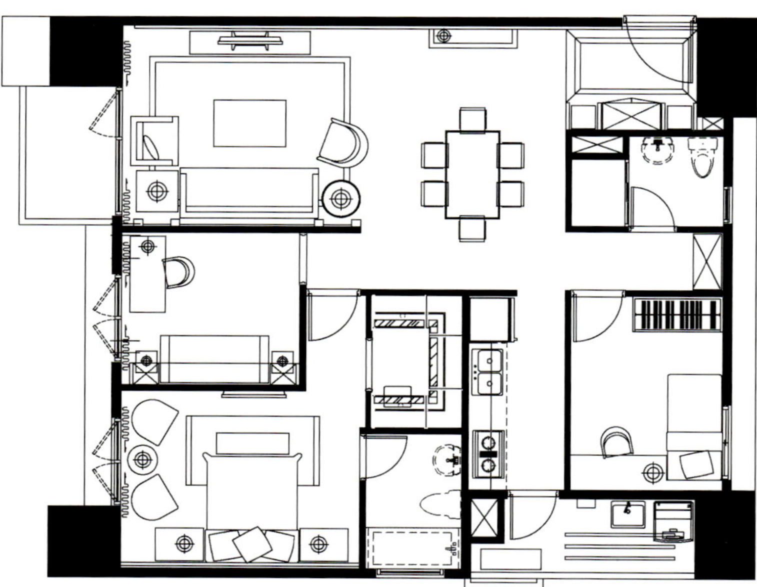

简约的灰色流光溢彩。民族风和时尚印花贯穿其中，一样也不肯落下。与客厅合并的餐厅比较好布置，这种格局的餐厅装饰性重于功能性，以美观为主。同时，这里的主空间是客厅，餐厅的装修格调必须与客厅统一，否则两个相连的空间会给人一种不协调的感觉。

黑与白的色彩搭配给人一种强有力的视觉冲击，大面积的黑色烤漆玻璃和黑色系的电视机，以及音响与浅色系形成鲜明的对比，给人一种典雅的时尚气息。米色的沙发搭配简单的小碎花靠垫，给人舒适温暖的感觉，简约中又不失温馨。白色系的主旋律具有强烈放射光线的能力，可以使房间看上去更明亮。

风格前卫，另类抢眼，或张扬或低调，都是在追求个性的与众不同，纯白色的台盆与马桶使这个小小的卫生间显得通彻明亮。

Farglory Xindu Show Flat

远雄新都样品屋

设计师：黄书恒　参与设计：欧阳毅、陈佳琪、王建益、胡春惠、胡春梅　设计单位：玄武设计（Sherwood Design）　项目地点：台湾台北　建筑面积：264平方米　主要材料：银狐石、白色冷烤漆、特殊壁纸、南非黑石材　摄影师：王基守

In this project, especially in the public spaces, the designers have chosen low-saturation design to bring out new classical atmosphere. Taking the living room for example, white fabric sofa and black master chair with white edging are processing a dialogue elegantly in the space.

Apart from that, to make the space less monotonous, the designers have placed two imported horsehair single chairs in the setting, which have maintained good texture, and also added the finishing touch.

linear panels and totem relief on the ceiling are allegorical symbols applied to the public space. In the dining room, linear panels are deliberately used to build up the round ceiling, while the center part is especially decorated with French-style relief to respond to the chandelier. The same relief pattern also appears on the square ceiling linear panel in the living room. The two relief patterns respond to each other, reflecting Sherwood Design' s careful intention of shaping a unified space.

On one side of the dining room, a socializing room that can also serve as a study room is reserved, which conveys peaceful, elegant and comfortable atmosphere, and also becomes a small world where the house owner can ponder all alone.

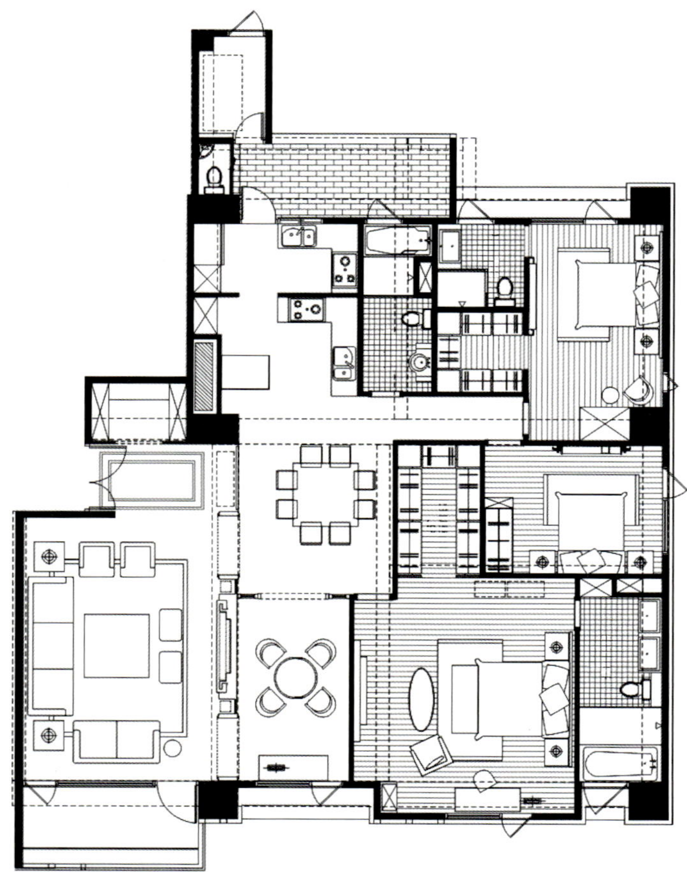

In the master bedroom, the designers still adopts the interlacing black and white to create a comfortable and slightly moderate style. The fireplace at the end of the bed hides the TV. A silver master chair with black flower pattern and satin gloss has lightened up the low-saturation space. It is worth mentioning that the dressing room of the master bedroom is specially designed by Sherwood Design into a show room like a high-class boutique shop, which enables the house owner to try on his/her best clothing as he/she like in the private space. Wall at the bed head is of simplified English classical style with exquisite symmetry of Baroque, but without heaviness of complicated engraving. The dressing table area has also maintained the consistent beauty that is brought out by the contrast between black and white. All the furnishings are carefully painted and trimmed to create low-profile exquisite style.

The design has chosen neo-classical style as the theme and simplified the complicated classical vocabulary, and further more, blended with the solemn atmosphere of English style, at last creating a simple and elegant neo-classical legend in this manor-like residence.

此案例设计师在公共空间中，特别以黑与白的低彩度设计，带出新调古典的气息。以客厅为例，白色布沙发与黑色镶白边的主人椅，在空间中优雅地进行着对话。

除此之外，设计师为了不使空间流于单调，遂将两把进口马毛单椅陈设其中，既带出质感，又起画龙点睛之效。

线板与天花板的浮雕图腾，也是此案例用以贯穿整个公共空间的寓意符号。餐厅刻意以多圈线板堆砌出圆形天花板，中央更以法式风格的浮雕与枝型吊灯相称。设计师更以此浮雕图案，特意同步呈现在客厅的方形天花线板上。两者互相呼应，更突显玄武设计师团队对塑造一贯风格的用心。

餐厅侧边更预留一处兼具书房功能的交谊室，除了传达静宁雅适的休闲气息，更为主人预留心灵沉淀的天地。

进入主卧室，设计师依然以黑白二色的交错变化，塑造舒适却又隐约自持的精致风格。床尾的壁炉造型，掩饰着电视摆放处，缎面光泽的银底黑花主人椅，点亮这一方低彩度的空间。值得一提的是主卧的更衣室，玄武设计团队刻意打造有如高档精品店的展示空间，让试衣的主人，也能尽情在这一方私密空间中旋转挥洒。床头壁板为简化的英式古典造型，有巴洛克的精细对称，却没有雕琢的繁复沉重。梳妆区也维持一贯的黑白色调对比带来的惊艳，家饰皆细心镶边，创造低调的精致风格。

选择新古典为主轴，将繁复的古典语汇简化，并融入英式的庄重气质，进而在这户庄园般的宅邸中，糅合出简约雅致的新古典传奇。

Wharf Holdings' Times Riverside Show Flat 30B
九龙仓时代尊邸30B示范单位

设计师：方峻（Tsun Fong） 设计单位：香港方黄建筑师事务所（Hong Kong Fong Wong Architects & Associates） 项目地点：四川成都 建筑面积：190平方米

The space is filled with the modern simple and elegant tonality of American-style with the color of off-white, showing the visual effect of the white board of gallery or showroom and setting off the display of the artist' s personal works of art and his collections with convenience to change them at any time. That is the home of a well-known contemporary Chinese American artist.

素净淡雅的米白色系，让空间拥有如画廊、展厅白色布景板般的视觉效果，映衬出室内陈设的艺术家的个人创作作品或收藏品，并方便随时更换。这便是一位美籍华裔当代知名艺术家的家。

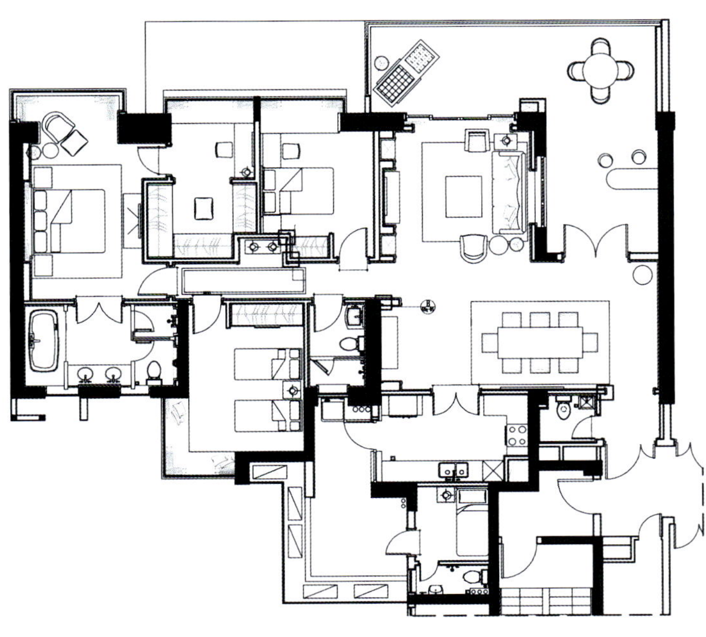

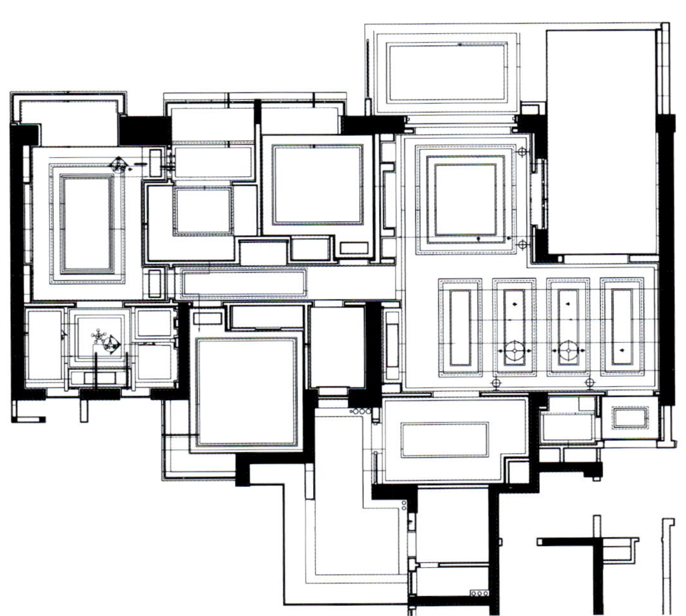

Ms. Xie's Residence in Shilin, Taipei
士林谢宅

设计师：张德良、殷崇渊　设计单位：演拓空间室内设计　项目地点：台湾台北　建筑面积：92平方米　主要材料：抛光石英砖、超耐磨地板、美耐板、积层木皮、密底板喷漆、茶镜、国产壁纸

The translucent glass vestibule replaces the original wall to bring the light in freely and to keep it private as well. The vestibule indicates the taste and individualism of the owner.

The sofa and furniture in drawing room are arranged in L shape according to the character of space. In particular the original partition of study room next to the drawing room is changed into the TV glass screen which can rotate 90°. It not only widens the space but also meets the needs for drawing room and dining room. Meanwhile the height difference of floor shows the regionality of the study room.

The good storage of dining room is very important for the house. The door of the storage cabinet is decorated with elements of Japanese architect Toyo Ito which breaks the rigid design. The very appropriate light adds much aesthetic atmosphere.

Besides, the passage to the room is too long. The door of main bedroom is designed into the secret door to widen the room so as to make up for the shortcoming of passage by transferring the space.

The main bedroom is simple and plain and the secret door connects the dressing room and washroom to maximize the use of space.

The moved interphone keeps the original manhole and circuitry for future maintaining, which is always ignored. We take every detail into our consideration such as partition of shoe cabinet to meet the needs of every family member.

半透明的玻璃玄关，让光线能够自由穿透又保证室内的私密性，也可借由玄关一窥屋主的品位及个性风格。

客厅沙发及家具则依照其空间特性做L形的摆设，特别是紧邻客厅的书房，经过屋主的同意后，改变原本用玻璃隔间做分隔的提案，改为可90°旋转的电视玻璃屏风，不仅开放的视觉效果让空间更开阔，提供客、餐厅两用的视觉功能，还以地板的高低差来表现书房的区域性。

餐厅良好的收纳功能，在家庭中占十分重要的地位。将收纳柜门的设计加入了日本建筑大师伊东丰雄的元素概念，突破原有刻板的门面印象，搭配适当的灯光照明，空间中充满了美学的气氛！

由于通往房间的走道过于狭长，因此将主卧房的门以暗门的手法做了适当的外推，让卧房空间变大，也利用空间转移的方式掩饰了这一不足！

主卧房的空间以简约大方为设计重点，并以暗门连接更衣间及卫浴间，使空间发挥更大的功效。

贴心规划：为移动过的对讲机，保留原检修孔及线路，方便日后维修；这是一般装潢施工常忽略的小细节，容易造成线路的混乱，维修不易，但我们的团队为您想到了！通过居家调查包括鞋柜隔层，都是需要考虑的内容，使每一位家中成员使用起来更方便。

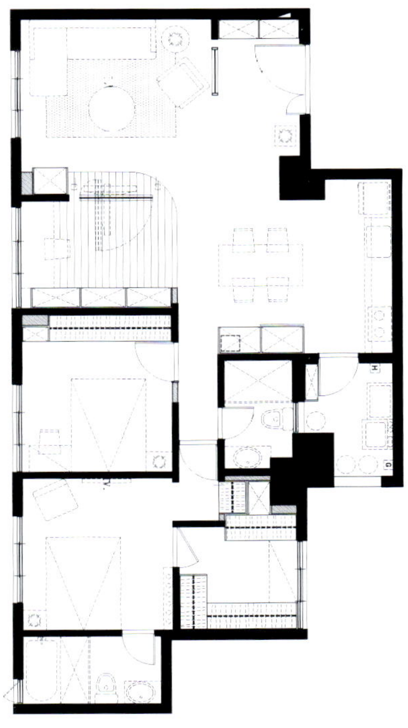

Pinzhendi Lin's Residence, Taichung
台中品臻邸林宅

设计师：黄金旭　参与设计：廖怡菁　设计单位：大言室内设计　项目地点：台湾台中　建筑面积：约331平方米　主要材料：大理石、文化石、水性木器漆、实木皮　摄影师：刘俊杰

The space is white toned. Simple and neat lines and no excessive furniture or decorations reveal a simple-style living space. Public spaces like living room and dining room are designed into a complete block. With the help of horizontal axis or vertical facades, the public spaces harmoniously complement each other in an independent yet connected way, which has made the space more spacious. Designers have planned out the configuration according to functions. The study room behind the dining room has adopted a sliding wood door. No matter the door is open or closed, the space can be enlarged in both ways. In one corner of the passage way stands a black stone sculpture of strange shape, which has added liveliness and energy to the space together with green potted plants.

Stepping into the daughter's room, you will see a completely white world which will make you mistakenly think of having come to a dream-like paradise. White bed, side cabinet, wardrobe and curtain are not only harmonious with the tone of the overall space, but also provide people clean and clear, cozy and comfortable feelings. Light blue backdrop makes the space fresher and more fantastic. In the master bedroom, the flooring has adopted warm wood material to create calm and comfortable atmosphere. Warm color and exquisite texture of wood material has toned down the contrasts between dark and light colors, made a proper transition and created warm and comfortable living atmosphere as well.

此案例以白色为主色调，简单利落的线条，无过多的家具与装饰，表现一种简约的家居生活环境。客餐厅等公共空间规划为一个完整区块，利用水平轴或垂直立面的呈现，在独立与连贯之间彼此呼应协调，延伸空间。设计师按功能区域划分空间，餐厅后以木质滑动门隔离的书房，合与关都有放大延伸空间的效果。在走道的一角，黑白木柱上竖立着一个造型怪异的黑色石塑，它与绿色盆栽都为空间增添了灵动与活力。

步入女孩房，满眼的白会让人误以为来到了梦幻的天堂，白色的床、边柜、衣柜、窗帘，不仅和整个空间色调协调，还给人干净剔透、温馨舒适的感觉。淡蓝色的背景墙使空间更为清新梦幻。在主卧部分，选用温暖的木质地板，营造出沉稳舒适的感觉。木质温润的色泽及细致的纹路，弱化了空间里大量的深浅对比，将深浅做了适度的过渡，也为居家营造温馨舒适的氛围。

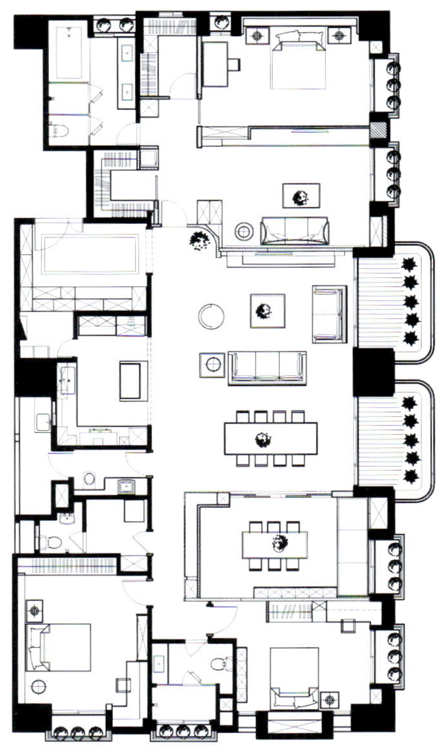

Aesthetics of Depth in Open Space

空间开阔的层次美学

设计师：唐忠汉　设计单位：近境制作设计有限公司　项目地点：台湾台北　建筑面积：130平方米　主要材料：洞石石材、银狐石材、不锈钢、玻璃、茶镜、柚木地板

The design focus in the space is a glass steel wall that starts from the entrance hall and turns into the interior space. As an important element in the space, the heavy and mottled stone used at the basis of the glass wall that goes through the entrance hall and the light, transparent iron glass steel material have made contrasts between lightness and heaviness, thinness and thickness.

Entering the space, you can see the transparent wall continues to stretch and becomes a partition in the kitchen. With glass and steel materials, the design has presented a multi-functional kitchen in a proper configuration. When the kitchen is changed into another focus in the living space, it has solved the smoke scattering problem that traditional open kitchen often faces, and also rearranged the kitchen space which is originally important, but often neglected.

空间的设计重点是一道从玄关一路转折进入室内的玻璃钢构墙面。作为室内空间的重要元素，穿过玄关的这一道玻璃墙面，底部主墙利用厚重斑驳的石材与前方轻透的铁件玻璃钢构形成材质上轻薄与厚实的对比。

进入室内后，墙面持续延伸，或为厨房空间的隔间，利用玻璃与钢构的材质设计，将一个功能完备的厨房空间“呈现”在合适的格局配置中，将厨房空间转换成生活空间的另一重心，不但解决了传统开放式中式厨房油烟四散的问题，并且将原本在生活中十分重要但却受到忽视的厨房空间进行重新的安排。

Long Space of Linear Structure

线条结构空间长卷

设计师：唐忠汉　设计单位：近境制作设计有限公司　项目地点：台湾台北　建筑面积：200平方米　主要材料：伯爵石、白橡木集成材、黑檀集成材、风化木皮、不锈钢、铁件、编织牛皮、特殊壁纸

A stone main wall stretching from the floor has partitioned the living room and the entrance hall. The wall is deliberately approached with origami concept and presented with different techniques, which has connected the left and right walls. Branch-shaped dimensional totem stretches through the center axis in the living and becomes a partition element of another space. Change in depth has added to the openness of the space and enlarged the space to the fullest. Dimensional texture of the materials makes the visual experience in the space a richest one. The living space is no longer a cold show, but has changed into a charming world that can record and collect life track.

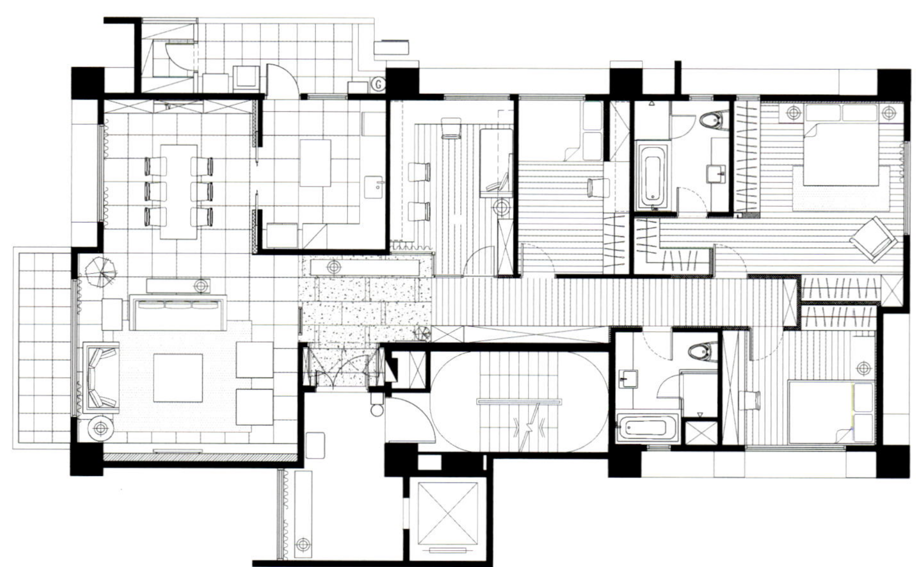

用一面从地面转折延伸的石材主墙，区分了客厅、玄关两空间，刻意将此石材墙面运用折纸概念，以不同工法呈现，连接左右墙面，以树枝状的图腾立体延伸贯穿居家中心线，成为空间中的另一个空间区分元素。穿透的层次变化，增加了空间场所的开阔性，使室内的空间层次发挥出最大的放大效果。立体材质造就的质感空间，赋予了居者最丰富的体验，居家不再是冰冷的展示，转换成记录收藏生活轨迹的迷人空间。

Aesthetics of Black of Deep Turn

转折深邃的黑色美学

设计师：唐忠汉　设计单位：近境制作设计有限公司　项目地点：台湾台北　建筑面积：165平方米　主要材料：黑云石、橡木染深、锈铜地砖、不锈钢、玻璃、铁件、茶镜

With the concept of interior architecture, the design has conveyed a continuous, transparent space with depth. TV wall as the main body in the space is connected with the shoe case in the entrance hall and forms the theme of interior architecture. Besides, the transparent glass partition walls together with the changes of floor tiles, present a transparent space with depth. Collocation of dark color materials reflects the deep aesthetics of modern humanities.

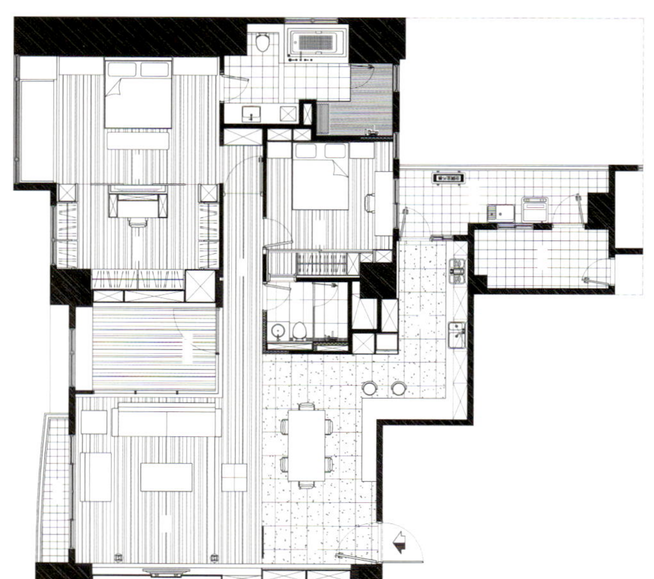

以室内建筑的设计概念，表达延续穿透、有层次的空间意境。将电视主墙视为空间的主体，连贯玄关鞋柜，形成室内建筑的主题。另外，以玻璃隔墙搭配地坪铺面的变化，呈现层次穿透的空间体验。以深色的材质搭配，表现现代人文的深沉美学。

Gorgeous Elements Deliver Romantic Charm

华丽元素传递浪漫情味

设计师：詹秉萘　设计单位：舍子美学设计　项目地点：台湾新北　建筑面积：122平方米　主要材料：天然石材、明镜、木皮、壁纸、玻璃、黑镜、板岩、花砖、风化木

Front yard wall is paved with natural stone, reflecting original rough feeling. It is a creative design to embed the number plate of their house in the form of a shop sign. Surface of the shoe case is a large full-length mirror, which has maintained practical function. The entrance has fully showed grand atmosphere, which makes people expect more to see the interior configuration.

In the living room, the TV wall is decorated with special teasel stretch fabric and mirror of modern luxurious mix-and-match style. Gentle curves of the white elegant custom-made sofa has softened the space and also created romantic and beautiful atmosphere. The raised study room is partitioned with sliding door from the dining room, where black mirror and veneer bring out exquisite texture. The dining room is of open style. The buffet is made of leather, veneer and glass, showing aesthetic beauty of contrasts of different materials.

Floor of the raised study is exquisitely paved with pipe lights and pebbles, which are full of leisure charm. A floor lamp at one corner of the dining room has lighted up the wall, whose splendid wallpaper with totem patterns is full of modern feeling. The other side is interspersed with decorative veneer and mirror, endowing the vertical surface with rich and varied expressions. Steady and gorgeous elements with rich texture are flowing in the space, giving off romantic charm through details.

The master bedroom has continued the splendid and modern style in public space. Carefully crafted door cleverly hides the bathroom, which is paved with slates and patterned piles, containing liveliness in elegance. Entering the space, people can remove precautions, relax completely and wash away tiredness of their bodies. The second bedroom takes light colors as the main tone, in which warm texture of wood seems to have the magic of soothing people. Stay in the quiet space and you can make up your spirit through sleep and get ready for all kinds of challenges. Designer Zhan Bingying is good at mix-and-match of a variety of materials to create stunning spatial atmosphere, once more showing terrific design skills.

前院墙面以天然石材铺设，展现原始粗犷的感觉，把自家门牌以招牌的造型镶嵌在墙上别具创意。鞋柜表面选用大面穿衣镜兼具实用功能。外玄关就充分展现出了大气风范，令人更加期待室内的空间规划。

走进客厅，电视墙选用特殊的毛草绷布和明镜，穿插混搭出时尚奢华感；白色优雅的定制沙发微弯的曲线，既柔和了空间也营造出浪漫唯美的氛围。架高的书房空间以拉门和餐厅作出分隔，运用黑镜和木皮带出精致质地；餐厅以开放式手法为空间定调，餐柜以皮革搭配木皮和玻璃，展现材质混搭的冲突美感。

架高书房下方以水管灯及鹅卵石细细铺陈，弥漫着休闲意趣。餐厅旁一隅以立灯打亮壁面，华丽的图腾壁纸饶富时尚感；另一侧木皮及明镜穿插点缀，让立面拥有丰富多变的表情。空间中流动着沉稳、富有质感的华丽元素，从细节传递出浪漫情境与韵味。

主卧室承袭公共领域的华丽和时尚，精心打造的门巧妙地将卫浴空间隐身，卫浴以板岩及花砖铺设，淡雅中富有活力。进入这个空间，人们可以卸下武装，完全地放松，洗去一身疲劳。次卧以淡色为主色调，温润的木质纹路似乎有种安抚人心的神奇魔力，徜徉在静谧的环境中，通过睡眠补足精神准备迎接各种挑战。设计师擅长运用多材质的混搭，制造出令人惊艳的氛围和空间感，再次展露深厚的设计功力。

Lee's Residence at Sanhe Street, Beitou

北投三合街李公馆

设计师：詹秉萦　设计单位：舍子美学设计　项目地点：台湾新北　建筑面积：132平方米　主要材料：复古砖、风化木、复古画框线板、壁纸

How to reconstruct the original commercial space of a kindergarten into a comfortable home space? The project possesses natural landscape and high ceiling penthouse configuration, which are rare in downtown area. Just open the window and you can overlook beautiful waters and mountains. Designer Zhan Bingying from Shezi Design confidently changed all the configuration and created elegant retro European-style space with natural materials and warm, elegant charm in south of France.

To maintain good interior lighting, the designer has specially chosen irregular shaped iron bars as handles on the stair, which plays as a partition in the living room together with clear tea-color glass with pattern, letting day light penetrate into the interior and avoiding the strong draught. In the living room and dining room, natural stone is the building material for half of the TV wall, which has stretched the vision.

In the dining room behind the TV wall, half of the wall is changed into a tea color glass with classical texture, which corresponds with the wallpaper. The beautiful landscape is not wasted at all. The designer intentionally added a table and chair beside the window, from where people can see the scene outside. How admirable it would be to simply take an elegant afternoon tea in the leisure corner besides the green view outside!

Along the clear stair, we come to the second floor. To create good atmosphere near the staircase is really important, so warm earth color wallpaper is chosen, on which indirect light shines, leading people into the warm space. Through the antique window, landscape at the end of the stair brings beautiful outdoor landscape into interior, which is natural and unaffected. Sofa seat at the corner again enables green view accessible at each corner, fully creating leisure atmosphere of south of France.

Behind the classical frosted glass sliding door is a flexible space, which on ordinary days can serve as house owner' s study room, and when guests come, they can sit on the ground casually. It would be nice no matter chatting or drinking tea, while accompanied by exterior landscape, leisure atmosphere filling the space.

When coming to the master bedroom, you will first see an out-pushing small space which is a dressing table reserved specially for the hostess. When the hostess finishes her making up, she can watch the scene outside the window as she likes and start a perfect day. The master bedroom is warm and stylish. The designer has decorated tea color mirror with line panel paint. One-piece-style storage cabinet makes the space not a little bit crowded.

In the penthouse which is rare in downtown, a 6-meter ceiling is maintained. Designer Zhan Bingying has not wasted the precious outdoor landscape. Every corner in the house has kept a small resting place for leisure. You can either brew a pot of good tea or a cup of fragrant coffee. Just enjoy your comfortable life at home now!

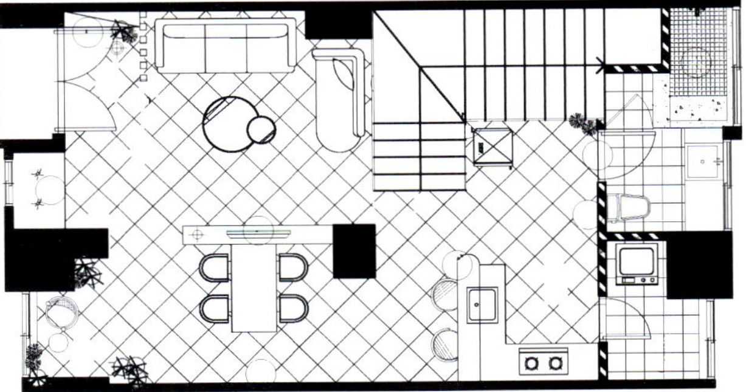

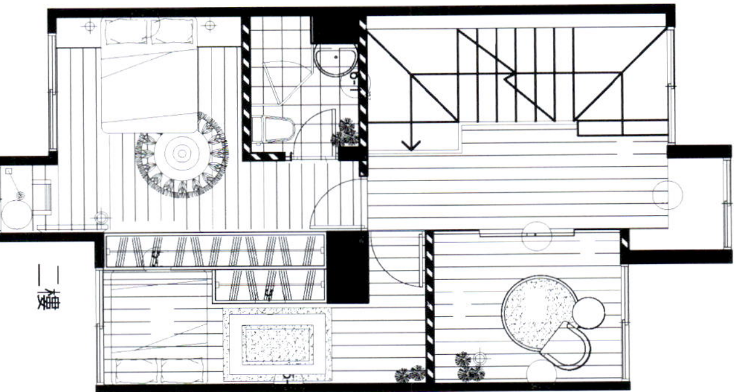

原是幼儿园的商业空间，要如何改装成为舒适的居家环境？此案例拥有市区难得的自然景观，挑高跃层格局。只要打开窗，好山好水尽收眼底。舍子设计的设计师大胆变更所有格局，以法国南部温暖优雅气质、天然素材，打造雅致复古欧洲情怀。

为了保证空间采光，设计师特别以不规则铁条作为楼梯把手，搭配清透喷图的茶色玻璃，让屋后光线得以穿透，也避掉穿堂煞风水弊端；客餐厅空间，以天然石材作为客厅电视背景墙面，该墙为半高的分隔墙，让视野得以展延。

在电视墙后方的餐厅空间，半高墙面则改以茶镜喷上古典纹路，与壁纸相互辉映。不浪费任何一处好视野，设计师特别在窗户旁增设桌椅，可在此欣赏户外景致，悠闲生活令人羡慕。

顺着楼梯进入二楼，楼梯旁的气氛营造当然不能放过，温润大地色壁纸打上间接灯光，引领进入温暖空间；楼梯端景通过复古窗台带进户外美景，自然不造作。角落旁的沙发座，更让绿意景致无死角，充分营造法国南部休闲风格。

古典雾面玻璃拉门后方，是能灵活运用的空间，平日可当屋主书房，有客人来访时，随意席地而坐，聊天泡茶配上窗外美景，优哉氛围自然而生。

进入主卧房，首先见到的是外推的小小空间，这是女主人专属的化妆台，化妆之余，偶尔望向窗外风光，完美一天就此展开；主卧空间温暖时尚，设计师以线板烤漆搭配茶镜，整片式的收纳柜体，让空间不显拥挤压迫。

市区少见的跃层空间，仍保有六米的挑高高度，设计师不浪费屋外难得的景致，在家中各角落保留悠闲休憩小区块。沏壶好茶、泡杯香浓咖啡，怡然自得的舒适生活，从家开始。

Shulin Spring World

树林泉世界

设计师：许宏彰　设计单位：德力设计　项目地点：台湾新北　建筑面积：99平方米　主要材料：石文木、古典胡桃山形纹、烟熏橡木地板、白砖墙、梦幻大理石、灰镜、烤漆玻璃、石英砖、铁件、结晶钢考、秋香木、白梣木

The owner is eager for an American mixed house. Therefore, the "cultural stone" imitating red bricks storage is the background and everything comes from this. Though it is mixed, all the materials and decorations, such as ceramic tiles, wooden and leather textures, wood floor, furniture as well as the colors, should be interlocked in order to create a free and easy house by finding the sameness among the difference or in an opposite way.

Several original walls are pulled down and the entrance of master room is changed to create a spacious space. The key point is to create the crossing of diagonal visions in order to maximize functions and stereo perception of space.

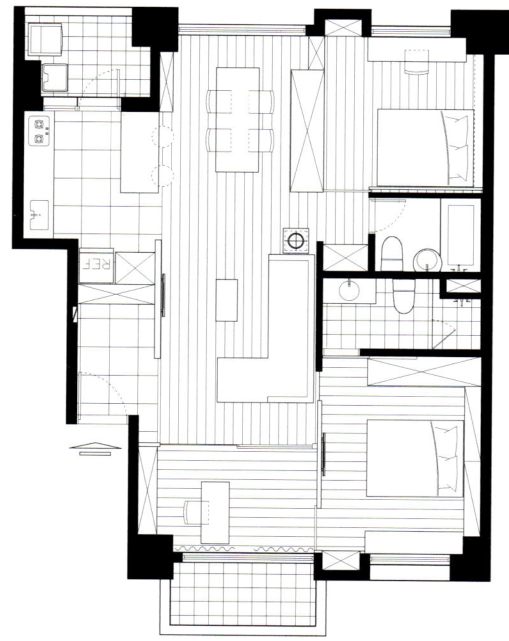

屋主一直期待拥有一个美式混搭风格的家。因此，仿红砖仓库的“文化石”便成了整个空间的基调，于是所有的都因此而生。虽说是混搭风格家居，但不论是瓷砖质感、木皮纹理、木制地板、软装家饰，乃至配色，仍是必须环环相扣，以达到异中求同、同中求异的自由与洒脱。

至于格局配置，设计团队说服屋主拆除了几堵关键的隔墙，并改变了主卧的出入口方位，营造一个尽可能开阔的空间，关键就在于创造对角视线的交叉穿透，让整个空间的功能与立体感都呈现最大化的可能。

swimming pools
Furniture

Ocean Star Residential Apartment

海华国际之星

设计师：许宏彰　设计单位：德力设计　项目地点：台湾中坜　建筑面积：112平方米　主要材料：金丝藤、黑檀、玉檀香、灰镜、烤漆玻璃、非洲铁木、砂岩砖

Designers make full use of existing wall to build cabinets, which are used to collect shoes, CD and DVD. Woody veneers with glass and grey mirror are applied to lead sunshine into the space and reflect shadow of living room to give a sense of hierarchy in vision. The visual background creates an unbroken space without visual disturbance from cabinets, which make people ignore the guest bathroom behind the wall.

Designers utilize various skills and approaches to change "constraint" to "characteristic" of the space, maximizing the value of design.

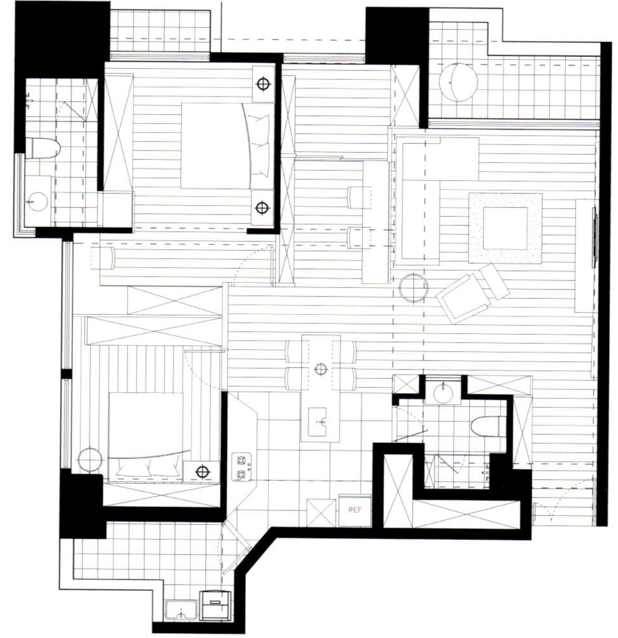

设计团队利用已有的墙面规划出鞋柜与收纳柜，更利用木作与贴皮技术，融合玻璃与茶镜，引导一部分光线进入室内，也可适度地折射客厅的倒影，在视觉上添加了细腻的层次感。整面的视觉主墙烘托出一个不受收纳柜干扰的完整空间，让人们忘记了隐藏在后面的客用卫浴间。

设计师利用各种巧妙的技巧与手法，将空间的“困境”转变为“特色”，实现设计价值的最大化。

The Age of Painting, Banqiao
板桥画世纪

设计师：许宏彰　设计单位：德力设计　项目地点：台湾台北　建筑面积：158平方米　主要材料：黑檀、玉檀香、灰镜、烤漆玻璃、抛光石英砖、河床石

The re-built house is designed to meet the needs of the owner and practical needs. The two original walls are removed, one of which is between the dining room and kitchen and is changed into bar counter separated from the electric appliance cabinet. The other original wall is between the secondary bedroom and drawing room and changed into glass sliding door which keeps both transparency and privacy.

Looking from dining room to drawing room, in the right side is the study; looking from bar counter to drawing room, the reflection of grey filters widens the space. The two pictures show the unique design idea for the drawing and kitchen. The small refrigerator and wine cabinet near the double sofa are put in the study room and main bedroom respectively. The dressing room with bi-parting door is at the back of bedstand in the main bedroom.

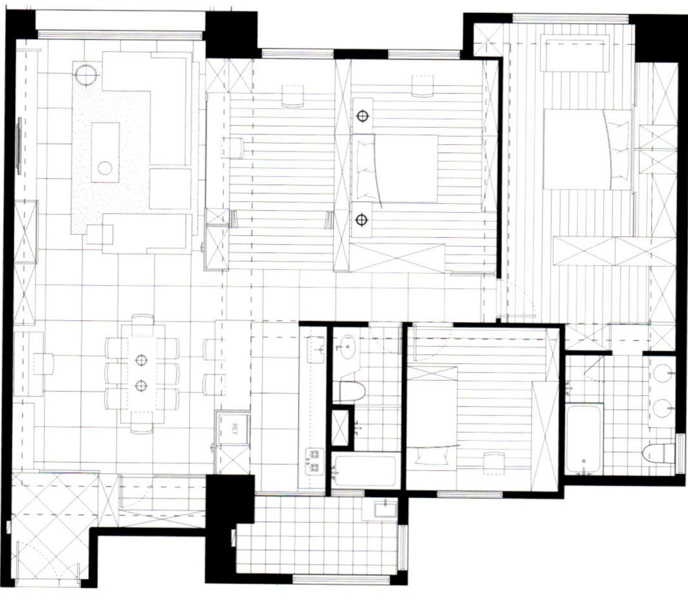

设计团队为了同时兼顾屋主偏爱的风格取向和实际需求，特别将两堵隔墙拆除：一堵位于餐厅与厨房之间，改以吧台与电器柜隔开；另一堵位于次卧与客厅之间，改以玻璃与拉门形式，在保留空间穿透性的同时，兼顾了使用上的私密性。

从餐厅望向客厅一景，右侧是书房。从厨房吧台望向客厅，设计团队运用了灰镜的反射特性，让空间更开阔。从客厅望向厨房，更加突显设计团队针对厨房与餐厅的设计做法。书房一景、主卧一处奇零空间放置小冰箱与小酒柜，旁边还可加一张双人沙发。主卧里的床头柜后面是一个双拉门更衣间。

He Huan Yujing

合环御景

设计师：许宏彰　设计单位：德力设计　项目地点：台湾台北　建筑面积：132平方米　主要材料：黑檀木、意大利仿洞石砖、意大利压纹砖、烟熏橡木地板、灰镜、烤漆玻璃、波斯灰大理石

The owner of this house is engaged in financial industry and has very rich experiences of top hotels, so he has a blueprint and spatial aesthetics of his home. After owning this house, the owner hopes that his future home can combine the elegant and leisure atmosphere and elements.

Guided by the logic of retreating in order to proceed, the design group arranges guest room between living room and balcony and keeps the living room with square configuration. By doing this, there's a transitional space from the outside balcony, and more scenery is added then. Besides, the wall between kitchen and dining room is destroyed and replaced by a bar table. A wine cabinet is beside, and then a relaxing and wine tasting space is created. One thing that should be emphasized is that the design group readjusts a little the relationship of master bedroom and its wash room, then a dressing table is involved in the design of wash room, so the area similar to exquisite hotel comes into being.

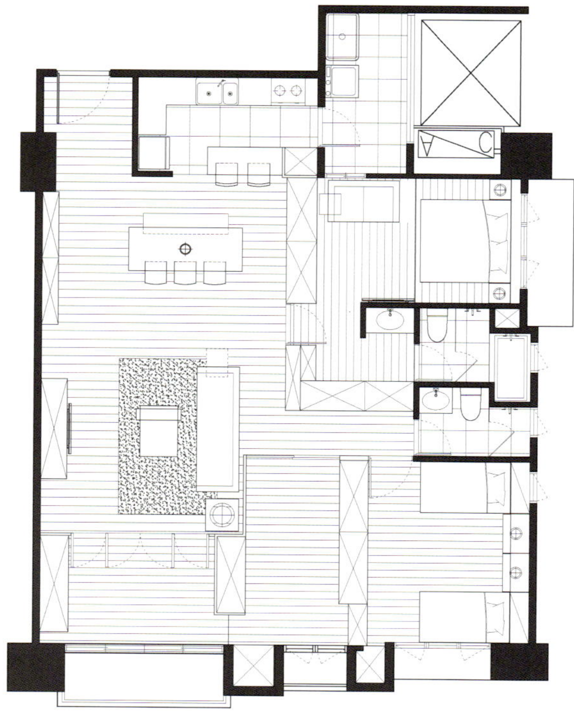

从事金融业的该户屋主常常旅游海外，有丰富的顶级旅店的空间体验，对于自家宅的空间美学可谓胸有成竹。因此，在购下此宅后，屋主希望未来的家也可以融入雅致闲适的空间氛围与元素。

设计团队运用以退为进的逻辑，在客厅与阳台之间规划了客房，保留方正格局的客厅，并借此与户外阳台多一道缓冲空间，也多一道风景。另外，打掉厨房与餐厅的隔墙，以吧台加以分隔，一旁设有酒柜，让生活多一处放松与品酒的空间。特别对主卧与主卧卫浴间进行了重新设计，将梳妆台纳入卫浴间，为住户提供了一个类似精品旅店的生活空间。

Cang Jin Ge

藏金阁

设计师：许宏彰　设计单位：德力设计　项目地点：台湾内湖　建筑面积：72平方米　主要材料：秋香木皮、橡木木皮、烤漆玻璃、柚木地板、绿板岩

This is an interesting case, and it' s not decoration of old house, and not a design for small room. In fact, it' s a small space made of two units. The design group makes the wall between two houses through, and only chooses one as the entrance and the other as kitchen. Besides, other space is arranged into entrance, living room, dining room, children' s room, master bedroom, laundry room and study also playing the role of guest room. Although it' s small, it' s functional. Then the big beam between two houses comes into being a difficulty. The designer adopts the way of arch cover combining with indirect lanterns, and turns it into a part of lightening of living room. Besides, the theme wall made of stones lets the small house be a grand space.

Seen from the living room, one balcony of the original two is still laundry, the other turns into dining room. At the entrance, the designer makes use of the differences between stone and wooden floor to let the difference of inner and outer spatial atmosphere stand out, obtaining the spatial hint of emotions and surroundings.

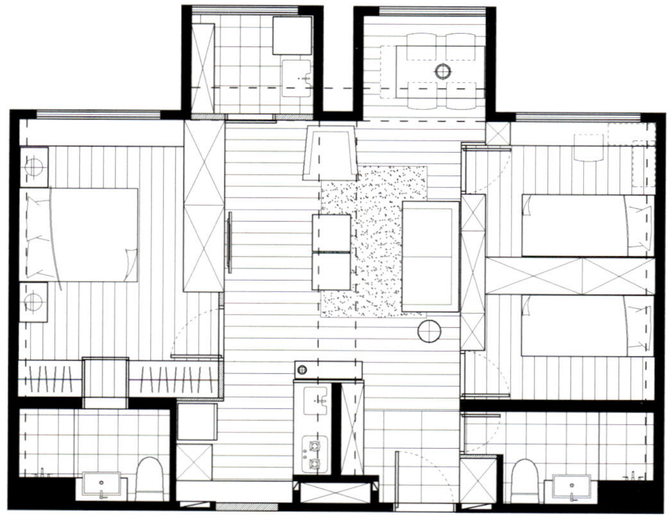

这一户相当有趣，它不是老房子装修，也不算小空间设计，而是两个单位拼在一起的小空间。设计团队打通两户间的隔墙，选择一个出入口，另一个封闭成厨房空间。此外，其他空间则规划成玄关、客厅、餐厅、儿童房、主卧、洗衣房以及一个书房兼客房，可谓麻雀虽小，五脏俱全。接下来的难题就是要化掉两户单位中间的大梁，设计团队采用圆弧包覆辅以间接灯的方式，使其变成客厅照明的一部分，并且使用石材铺砌成为空间的主题墙，让这个寸土寸金的小套房瞬间变得落落大方。

从客厅望向阳台，两户的阳台，一个仍是阳台（洗衣房），另一个则成了餐厅。在玄关，设计团队利用石材与木地板的材质差异，突显空间内外动静的氛围落差，达到情绪与情境上的空间暗示。

Wenshan Travelling House

文山行旅住家

设计师：许宏彰　设计单位：德力设计　项目地点：台湾台北　建筑面积：73平方米　主要材料：石文木、古典胡桃山形纹、秋香木、白梣木、黑色半抛光石英砖、梦幻大理石、灰镜、烤漆玻璃、抛光石英砖、铁件、紫檀木地板

Though it is the small family the owner still asks for a flexible study room which is used as drawing room when friends come. The baby room takes advantage of the height to design a special structure "lower is the study desk and upper is the bed".

The clean glass and grey mirror are used to connect the drawing room, kitchen and dining room to create circle structure and to keep them half-open in vision. You can see the drawing room and guest room from the kitchen and can also see the kitchen and drawing room in the guest room. Therefore the light is totally free and sublimates the whole space.

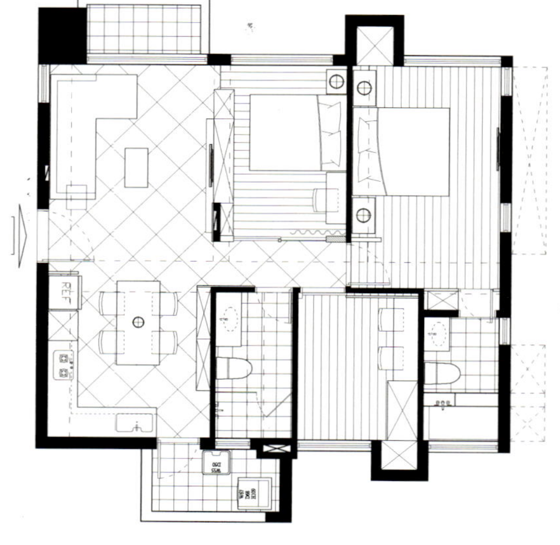

尽管此户的居住人口偏少，屋主仍要求有一间可弹性使用的书房兼客房，方便亲友入住。因为此户型挑高极高，儿童房便充分利用挑高，另增一“下为书桌上为睡铺”的复层。

特别值得一提的是，为了创造客厅与厨房兼餐厅以及客房三个各自独立区域之间的联系，特使用清玻璃与灰镜，营造一个“回”字形的视觉动线，让三个空间不再是彼此封闭，而是拥有着视觉流动性的半敞开空间。在厨房可以看到客厅与客房，在客房可以看到厨房与客厅，甚至是一部分的阳台风光。如此，不仅光线可以自由地穿梭，而且空间得到蜕变和升华。

FREER

Hua Jing Yuan

华景园

设计师：许宏彰　设计单位：德力设计　项目地点：台湾天母　建筑面积：138平方米　主要材料：栓木皮、胡桃木、烤漆玻璃、集成柚木地板

Nordic design combines modern simplicity and humanism, warm and nice. This style is a home decoration preferred by a certain people. Firstly, the design group evaluates and regroups the light and moving lines of space, hoping the freedom can flow in the space. Therefore, the design group plans to destroy the walls that don' t affect the secure structure, for example, the wall between living room and kitchen is dealt with to create an open and echoing space. The male owner can enjoy the freedom of coming and going between study and living room, and female owner can cook and take care of children in the kitchen and dining room at the same time. In fact, the action of rearrangement has a most important aim to get rid of the door facing the people who just enter.

The materials adopted are mostly with the preconditions of brightness and reducing the visual pressure to the utmost. Meanwhile, avoid all possible decorative fabrics that may cause mites that the female owner' s worried about since the home should be comfortable and safe for newborn baby.

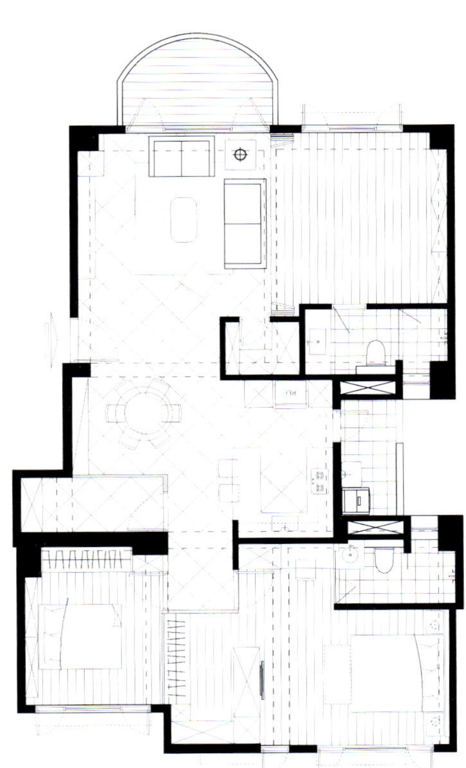

北欧设计糅合了现代简约与人文素养，简单而隽永，温润而有型，这样的风格也是个人相当偏爱的居家风尚。首先，设计团队针对空间的光线与动线进行评估与整合，希望让空间具有流动的自由气息。因此，设计团队决定打掉不影响结构安全的隔墙，诸如客厅与厨房，营造一个开阔无碍、遥相呼应的空间，男主人可以在书房与客厅间自由穿梭，女主人可以在开放式的厨房与餐厅空间一边烹调一边照料孩子。其实，这些格局的调整，有一个最重要的目标就是化解一入此宅即面对客用卫浴门的情况。

材质的应用多以明亮为前提，尽可能将视觉压力降至最低。同时，回避所有可能造成女主人困扰的家饰布，以提供给新生儿一个舒适又安全的家。

Butterfly's Dance House

蝶舞行馆

设计师：许宏彰　设计单位：德力设计　项目地点：台湾

The owner is a loyal Buddhist, so his belief can be reflected from the life. Simple, humane, pure, comfortable and cozy spatial characters can really show the expectation and personality of the owner. Therefore, the designers adopt penetration principle to resort the facilities. Firstly, two walls are destroyed, and glass is used to separate the study and living room which can complement each other. Then the kitchen and dining room are connected by the bar table.

In order to let space have more layers, and more transfer from outdoor to indoor, the designers use pebbles partially to make room attract attention. The design above the TV cabinet is without frame. Getting rid of unnecessary edges can make room open.

The owner insists on keeping attached utensils. And with the precondition of not destroying utensils, break a wall to allow the possibility of interaction in space by the way of bar table. Replacing the wall between study and living room by tempered glass can extend the space. Designers replace the wall of the washroom of master bedroom with sliding door to make good use of space.

屋主是虔诚的佛教徒，因此要将其个人信仰反映在生活中。简单的、人文的、质朴的、舒适的、安静的空间特质，最能映照出屋主的内在期待与特质。正因如此，设计团队运用穿透原则进行配置重整，首先打掉两堵墙，让书房与客厅中间以玻璃分隔，两个空间交相呼应，如此也使厨房与餐厅借助吧台联系在一起。

为了让空间更有层次，让户外与室内之间多一份情绪上的转换，设计团队在玄关地板与吧台下方局部采用了河床石，让空间更具焦点。电视柜上方的无边界设计，化解掉非必要的棱角，可让空间更加开阔。

屋主坚持保留附加的厨具，在不破坏厨具的前提下打掉一堵隔墙，用吧台设计让空间与空间之间多一份互动的新可能；打掉一堵介于书房与客厅的隔墙，以强化玻璃替代，增加空间的延展性。为了更有效地利用每一寸空间，设计团队变更了主卧卫浴的隔墙，改以拉门，使整体空间更加通透。

Precise Techniques Create A Charming Space

精准工艺凝聚空间魅力

设计师：许炜杰Janus 设计单位：台北拾雅客空间设计 建筑面积：约50平方米 项目地点：台湾新北 主要材料：天然石材、不锈钢、木作喷漆、超白玻璃、进口窗帘、瑞士进口超耐磨地板、染白蜥蜴皮、期货家具

The most attracting part of this design probably is that it has carefully explored different features of the spaces and further more showed unique spatial characteristics. For every person, home is a place for collection feelings and memories, an area to take a rest and cultivate the mind. These demands have nothing to do with the location and the size of the house. The biggest concern of Shiyake Design team is to take people as the most important part in the space and make a harmonious dialogue between people and the space according to users' practical needs and preferences.

用心探索不同的空间属性，进而展现出独一无二的空间特质，是这个设计案最吸引人的魅力所在。家对每一个人来说，是用来收藏情感和记忆的地方，是休憩、涵养心灵的场所，这些与坐落位置、面积大小等附加条件无关。拾雅客设计团队以人为主体，依使用者的实际需求与喜好，搭建人与空间的和谐对话。

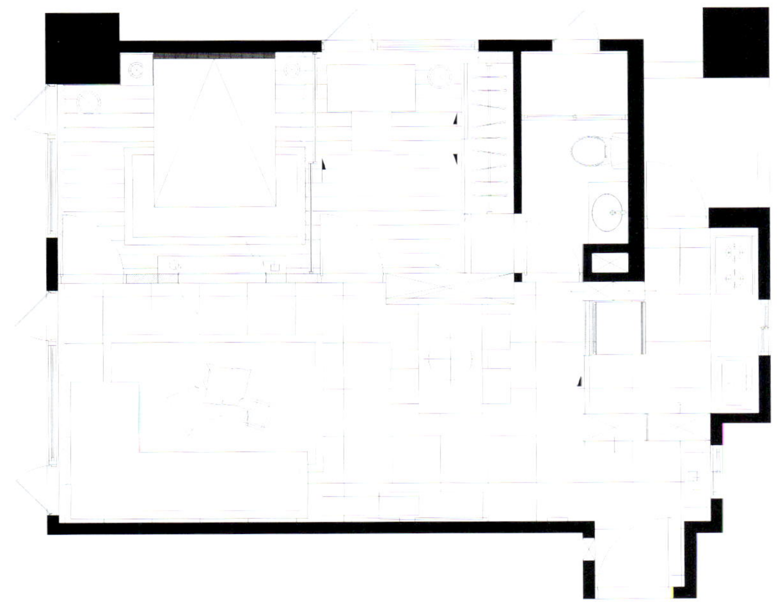

Water Park

水公园

设计师：许炜杰Janus　设计单位：台北拾雅客空间设计　项目地点：台湾新北　建筑面积：约188平方米　主要材料：线板、喷砂玻璃、实木柚木地板、镜面

Having traveled and lived abroad for years, the house owner is particularly fond of French romance and elegance. Designer Xu Weijie of Shiyake Design combines floral and natural elements which are the owner's favorites together into a romantic and leisure French living space.

Flying butterflies on the half-transparent frosted partition screen is the first thing to see when stepping into the space. You can follow the guidance of the flying butterflies on the flooring to get into the interior. In the clean white space, navy blue and orange sofa is colorful decoration that designer Xu Weijie chose for the mix-and-match solid teak flooring. Complicated line panel design has endowed the space with classical expressions. Line panel engraved on the wall and ceiling fill the space with aesthetic atmosphere. Butterflies fly past the window in the living room and the ceiling and mirror wall shared by the corridor and the study room, and finally rest on the frosted wall in the master bedroom to return to a quiet and relaxing state.

The house owner's collections around the world are presented in the display cabinets like the TV cabinet, buffet and bar counter. With the help of light design, the valuable collections are highlighted. A considerable number of mirrors are decorations in the space, which have removed the heaviness of the space made up of complicated lines and created a modern fashionable French living space.

长年旅居国外的屋主特别喜爱法国的浪漫优雅，因此设计师就以屋主喜爱的花艺、自然元素，串联起一个浪漫休闲的法式生活空间。

展翅的蝴蝶翩飞在半透明的喷砂格屏上，是踏入室内空间的第一视景，循着地坪上舞蝶的指引进入室内空间。洁净的白色空间里，宝蓝色及橘红色沙发是设计师混搭实木柚木地板的调色剂，繁复的线板设计搭建出空间的古典表情，浮刻于墙面及天花的造型线板则营造出唯美氛围。经过客厅窗边，蝴蝶飞过与廊道共构的书房天花及镜面墙，最后停歇于主卧室的喷砂墙面上。

将屋主旅行中收集的世界各地的收藏品，置于电视柜、餐具柜、吧台，以及搭配灯光设计的精品展示柜，更加烘托出珍藏品的不菲价值，通过大量运用镜面点缀在室内空间的手法，弱化繁复线条空间的繁重感，打造出现代时尚的法式悠闲居家。

Chen's Residence

三峡陈公馆

设计师：许炜杰　设计单位：拾雅客空间设计事务所　项目地点：台湾新北　建筑面积：149平方米　主要材料：黑檀木、枫木鸟眼、安丽格、巴拉圭紫檀木地板、不锈钢、茶镜、茶玻、灰玻、赛丽石、亚克力、人造石

The space is of American style with low profile layout to present a calm vibe. A spacious and leisurely space is needed by house owner who has been living in America for a long time, hence designer plans the open kitchen and dining room specially. Keeping in line with the airy atmosphere which is brought from folding door in reading room, the bar counter is designed to divide living room and kitchen. Half of the original layout is used as public area, which could not only bring the outdoor scenery inside, but also could accommodate many visitors. Main wall in living room uses characters of tawny glass and smoke glass to realize an airy and roomy visual effect, to accommodate the needs of rich color, to match the sense of space and material features perfectly in hierarchical arrangement.

此案例空间凭借着天然的材质刻画出美式的悠然，低调而内敛的格局，延伸出开阔从容的氛围。由于屋主长年居住于美国，偏好宽敞悠闲的空间氛围，因此设计师特别规划了开放式的客厅及餐厅。与书房折门所带来的通透感相互呼应，吧台被设计成为厨房与客厅的分界线。原始格局的二分之一被用做公共空间，不仅可以将室外风景揽入室内，也可以满足多位友人同时来访。客厅主墙利用茶镜、茶玻等材质的特性，实现通透、宽敞的视觉效果，满足了色彩的丰富性，在层次的排列上，完美结合了空间感与材质特性。

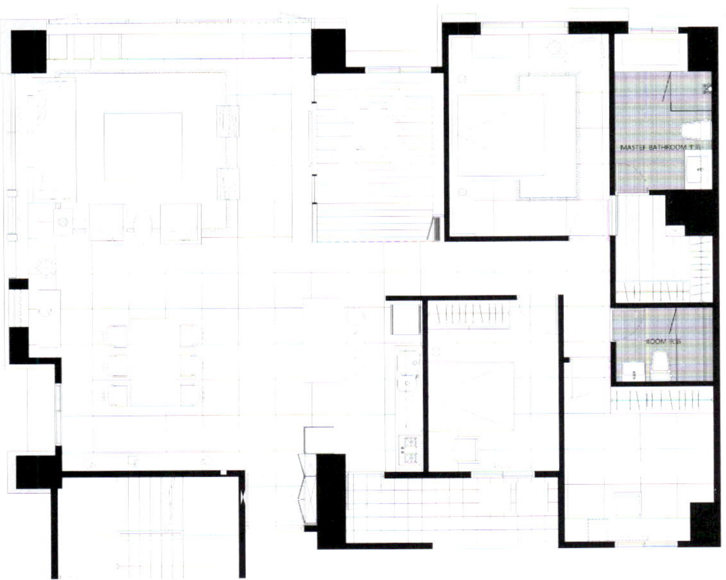

Gently Compose Music of Comfortable City Life

轻谱都市里自在的生活乐章

设计师：邱振民　设计单位：邱诚室内设计公司　项目地点：台湾台北　建筑面积：74平方米　主要材料：复古石英砖、镀钛金属板、风化木、铁刀木、铁件、裱布、进口马赛克意大利瓷砖、特殊色石英砖

Weathering wood in light color appears on the facade as the protagonist from the vestibule, which, in the match of the quartz flooring with the effect of the slate, constructs a leisurely space. The TV wall decorated with the titanium metal plate, the audio-visual machine in the bottom, neat geometric lines and cold materials neutralize the warmth in the living room. And the arc sofa in contrast avoids the traditional decorative techniques but transforms the design art of the public space into a gathering strength of stylish atmosphere.

The weathering wooden facade in light color around the whole house has the storage function of accommodating all the odds and ends of life. For the red mounted cloth often used in the bedroom, the designer adopts the cut-up technique to make it in scattered match with white iron furnishings so as to link the living room, the dining room and the kitchen. In a space with simple lines, the virtual door for entering the private field with patterns of old buildings on the surface enriches the decorations of the space just as the finishing touch and also brings out the casual and elegant life flavor.

Each room of the living field has plenty of natural light. When entering the private field behind the copper mesh, the open book shelf along the corridor in white lacquer coating comes before us. The shelf incorporates the owner's rich collection. The designer uses the guiding design approach to bring out the study in an invisible way. Under the shining of the setting sun, the window with a low frame and red brick walls with coated cylinder outline a beautiful perspective beside the window, reflecting a better life with the scenery in everywhere.

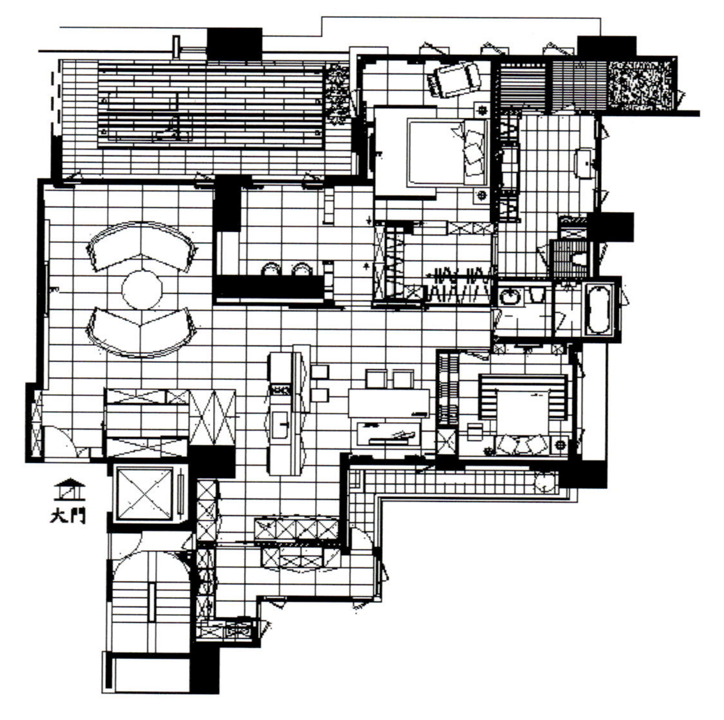
大門

在设计之初，屋主即表示不喜欢冷白的白色墙面，希望能在利落的空间线条里，打造有暖度的生活空间。浅色风化木从玄关入口处，即以主角之姿跃然空间立面，在类板岩效果的复古石英砖地坪搭配下，建构出闲适自在的空间基础，纵长的镀钛金属板电视墙与下方的视听机盒，利落的几何线条与刚冷的质材表现，中和了客厅里的暖度，而两两相衬的圆弧沙发，也跳脱传统摆设手法的框架，将公共空间的设计艺术，转化成为时尚氛围的凝聚力量。

环绕全屋的浅色风化木立面里，虚实掩映的线条后，有可容纳一整屋生活什物的收纳功能，常用于卧房空间的赭红色裱布，也被设计师以同样的切割手法，与白色铁件错落搭配，串联客厅与餐厨空间，虚化进入私人空间的门片、布面上斑斓的老建筑图纹，在线条简单的空间中，丰富空间层次，更带出随性的生活味道。

经过再次变化的生活空间，每个空间都拥有充足的自然光照，进入铜金属网门片后的私人空间，沿着廊道设计的白色烤漆开放式书架，轻易收纳屋主的丰富藏书，设计师引导式的设计手法，不着痕迹地带出书房空间的存在，压低窗框高度的窗户设计，与包覆柱体的红砖墙面，斜阳轻洒下，在窗边勾勒出写意的端景视野，体现随处即景的美好生活。

Luxury Dubai Style, North African Charm

迪拜奢华基调，勾勒北非迷情

设计师：邱振民　设计单位：邱诚室内设计公司　建筑面积：约661平方米　主要材料：黄金米色大理石、黑白大理石马赛克、黑云石、黑檀木、金箔、古董水晶灯

Light defines the existence of colors and also makes invisible soul possible. Under the light and shadow of classical chandelier light, designer Qiu Zhenmin has illustrated the brilliance of Dubai in the interior space with lines as painting brush, supplemented by luxury of gold foil and romantic classical French style decorations. Curved lines reveal a touch of Middle East charm and initiate the house owner's aristocrat-like exotic dreams.

The duplex villa of good location looks grand with its four floors, which have created breathtaking luxury and brilliance. Progressive gold foil door frame changes the flow of light and shadow. Rhythm of flowing lines of black and white marble mosaic in the splendid space has softened the toughness of the frame lines. With the decoration of large piece of oil painting facade, it has added into the space some cultural atmosphere.

After seeing the splendid and fantastic space, we come into the living room, where the 3.2-meter high ceiling and large suede sofa immediately tell about classical European-style vocabulary. With the help of Middle East totem carpet, classical style and folk custom have made the first contact. To blur the obstruction brought by Greek pillars, with carved paintings, designer Qiu Zhenmin tries to make visitors' vision refocused. Handrail of the stair that leads to the upstairs continues the classical style in the living room. Geometric patterns made out of bended iron and gold beige marble wind spirally up infinite depth, making people interact with the space.

Step on the stair and you will be greeted by the dining area. Designer Qiu Zhenmin has highlighted the texture of the stone flooring with another kind of simple style. Large curtains set a theater-like background for the curio cabinet. After a turn, we come to the long corridor. Symmetrical wood color door frames reflect the richness and warmth of the corridor. Complemented by indirect light source and collection show, they better highlight the modest luxury of gold foil landscape wall.

In the private area, French clinch wall at the bedhead in the master bedroom meets with the Middle East style bedhead again, which presents Morocco' s royal family-like bedroom. Soft golden tone space shows luxury of six-star hotel when furniture like bed end stool and bedside sofa is put into the space.

Exotic expressions added upon the luxury basis make every side of the villa possess extraordinary charm, which is Qiu Zhenmin' s design art, and another kind of enjoyment of luxury as well.

光线界定了色彩的存在，也成就了无形的灵魂。在古典水晶灯的光影下，设计师以线条为画笔，辅以金箔奢华、法式古典浪漫，将迪拜辉煌描绘入室，曲线下淡淡的中东情调，开启业主犹如贵族般的异国遥想。

走进精华地段的别墅内，四层楼双拼气势，构建出令人屏息的奢华大气，渐进式金箔门框转换光影流动，璀璨中映入黑白大理石马赛克的流线律动，软化了框线的刚硬感，加以大幅油画立面装点，增添一股人文气息。

客厅起居间，挑高3.2米，绒面大沙发佐以中东图腾地毯，古典与民俗风情有了亲密接触，且为了虚化空间内希腊柱体带来的滞碍，设计师以雕饰刻画，让来客视觉重新聚焦。通往楼上的楼梯扶手延续着客厅的古典风格，铸铁弯曲而出的几何图形，衬上黄金米色大理石，螺旋蜿蜒出无限景深，让人与空间开始有了互动。

拾阶而上，映入眼帘的用餐区域，设计师以另一种简约，突出地坪质感与石材纹理。通过大型窗布铺陈，古董柜有了剧场般的景深。转折后纵向的长廊上，木色门框为导引，体现出廊道的饱满温润，加以间接光源与珍藏展示，更加突显金箔端景的低调奢华。

在私密领域上，法式钉扣的主卧床头主墙，再次遇上中东风情床头，呈现摩洛哥皇族般的御用寝殿。柔和的金黄色，加之床尾凳和床边沙发家具，表现出六星级酒店般的尊贵。

在奢华基底上加以异国表情，让豪宅的一颦一笑开始有了不凡韵味，是设计师的设计艺术，也是进阶奢华的另一种享受。

Huang's Residence

黄公馆

设计师：邱振民　设计单位：邱诚室内设计公司　建筑面积：约132平方米　项目地点：台湾台北　主要材料：橡木地板、南方松、花墙板

Through the French windows sunshine comes into the room. What initially comes into view is simple pure white. When coming home, the couple is greeted by such simple but warm scene. There are no complicated decorations, no tension, only the most natural relaxing experience. To receive day light from both sides in the living room, the priority should be given to the control of the proportion of each space. Designer has dismantled the partition wall of the kitchen and guided day light on both sides in. Non-blocking transparent experience makes the space closer to the couple's expectation of a free and easy home life.

As to a beam running through the space, Designer Qiu Zhenmin did not treat the beam with conventional coating material or approaches like lowering the height. Instead, he made "light" meet with the beam to produce the illusion of an integrative design. Dining room embodies the overall simple culture and fills dinner date with quiet and comfortable atmosphere. The neighboring kitchen has left out concrete wall for hard division, so cooking will no longer be limited in a narrow and repressing space.

Wall panels with flower patterns are the best company in the lengthy passage that leads to private area. With the premise of a small family, only one room is kept as the master bedroom in the three bedrooms. The original dormer design can perfectly invite more sunshine into the master bedroom. Special wave shaped ceiling illustrates one and another beautiful arc in the sweet pink world and starts the romantic overture.

The terrace area is inspired by "the trance pavilion" where southern pine and plants bring comfortable atmosphere of nature. A deepened couch provides the couple a moment of leisure, on which they can either sit or sleep, or reading or even doing nothing. When they lie on the couch, it seems that they are in the hug of nature and look up to the blue sky, relaxing the constraints on the mind.

Designer Qiu Zhenmin has presented a natural and unsophisticated living space with simple design vocabulary. He blended the couple's favorites into the design and made a leisure living that the couple expected.

透过落地窗，日光洒落入室，最初映入眼帘的是简单纯白，回家时这样简单却温馨的场景，少了繁杂装饰，少了紧张感，留下最自然的轻松体验。为实现客厅的双面采光，在掌握各空间定位的前提下，拆除厨房的隔墙，将双面日光引导入内，无阻隔的通透感受，更贴近夫妻俩对于无拘束居家生活的向往。

贯穿空间的大梁，设计师不以一般包覆材质或压低高度的修饰手法，改以"光线"与之交会，产生一体造型的错觉。移步餐厅，承载整体的简单文化，让静谧自在的氛围填满晚餐约会；邻近的厨房区域，省略硬性区划的水泥墙，烹饪时不再委屈于狭窄压迫的小空间。

通往私人区域的冗长过道，花墙板装饰成为最佳的陪伴。在简单人口构成的前提之下，三间卧室仅保留一间为主卧房，原有的天窗设计，正好使主卧房沐浴在更多的日光之下。特殊的波浪天花造型，在甜美的粉红色国度，描绘出一个又一个的优美圆弧，浪漫序曲就此揭开。

以"发呆亭"为原型，休闲的露台区域，南方松与植栽花草带来惬意的自然气息，加深的卧榻，让两人偷一刻清闲，或坐或卧、放空或阅读，躺卧时，仿佛是赖在自然的怀抱望向蓝天，卸除心神上的束缚。

设计师以简单语汇，呈现自然不刻意的家居之美，将夫妻俩的爱好融入设计，营造出心驰神往的写意生活。

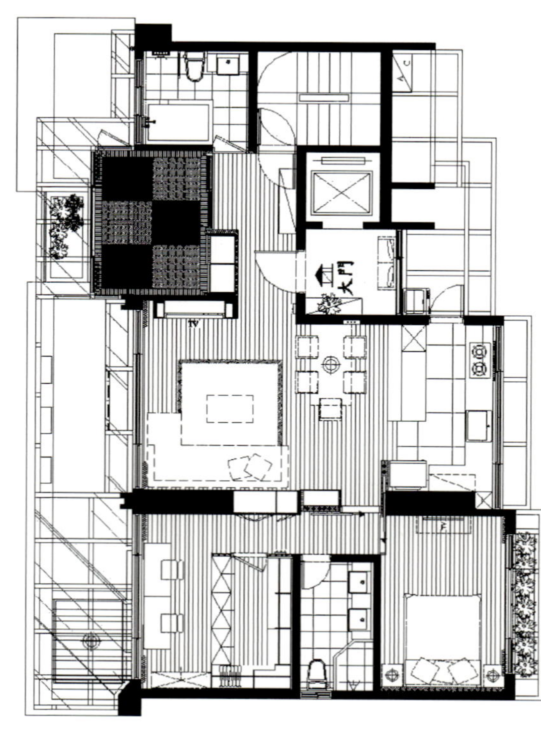

Chen's Residence, Tamshui

淡水陈公馆

设计师：张德良、殷崇渊　设计单位：演拓空间室内设计　项目地点：台湾台北　建筑面积：116平方米　主要材料：落地百叶、人造木皮、人造石、抛光石英砖、耐磨地板、镀钛铝板　摄影师：游宏祥

The project is a continuation of the designers' consistent modern design. The design has no excessively luxury decorations, but fits for modern people's life attitude which is simplified but not simple, plain but not dull, to possess a neat, comfortable living space where they can relax in the intense life pace.

The entire living room is connected with the dining room. Complicated configuration is abandoned, making the room spacious and grand. Good natural lighting condition matching pure white tone makes the whole space elegant and bright. With partly dotted with brown and black colors, the space shows a beautiful look. Smooth lines of the chandelier, door and windows, unique embedded door carved with rose patterns infuse the space with special, fresh atmosphere.

It is worth mentioning that when modernity is maximized, the designers have also added small amount of fresh pastoral-style elements. Rattan chairs, wood dining table and purple wallpaper in the bedroom create rustic atmosphere. Even in modern life we can have quiet, cozy and comfortable living space. The design of the bathroom is ingenious. A huge round bathtub and a transparent French window make up a perfect match. With the help of the beautiful landscape outside the window, people can relax their body and heart completely, when green plants come into sight, people becomes happy spontaneously.

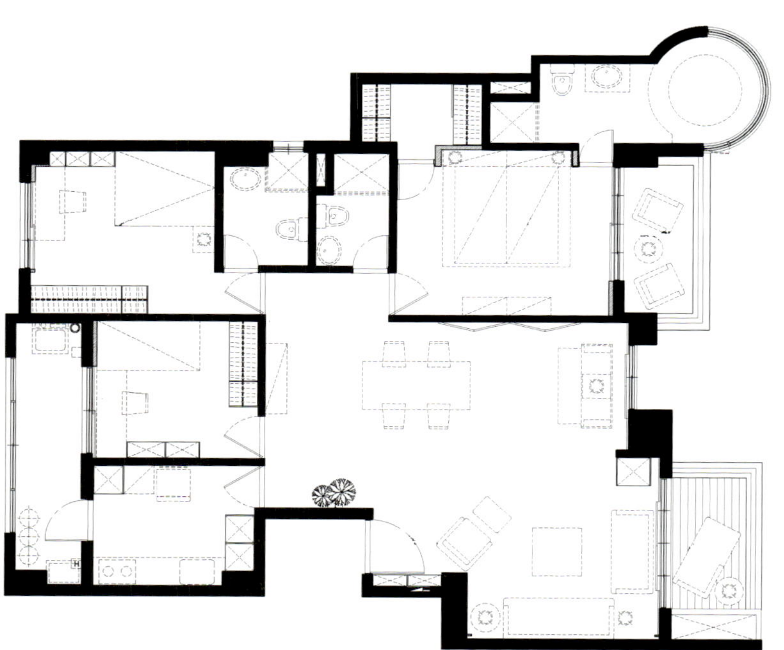

此案例延续了设计师一贯的现代风格，没有过分华丽的装饰，却契合了现代人对生活的态度——简约而不简单，平淡而不单调，可在紧张的生活节奏中拥有一个轻松简洁、放松心情的居家环境。

整个客厅与餐厅相连，摒弃了复杂的空间分割，宽阔大气，良好的采光配合着纯净的白色主色调，使整个室内都变得典雅明净，再佐以棕色、黑色的点缀，体现出空间“浓妆淡抹总相宜”的气质。吊灯以及门窗流畅的线条，别致的嵌入式玫瑰雕花门，都为室内注入了独特的新鲜气息。

值得一提的是，设计中还加上了些许清新的田园风格元素：藤编的靠椅、木制的餐桌、卧室紫色的墙纸，扑面而来的是回归自然的朴实气息。别出心裁的是卫生间的设计，一个偌大的圆形浴缸，一扇透亮的落地窗户，两者相得益彰。借助这窗外的美景，在完全放松身心的时候满是青翠，不禁使人的心情变得愉悦起来。

Lin's Residence, Sanxia

三峡林公馆

设计师：张德良、殷崇渊 设计单位：演拓空间室内设计 项目地点：台湾台北 建筑面积：128平方米 主要材料：人造木皮、抛光石英砖、茶镜、黑镜、耐磨地板 摄影师：游宏祥

A nice day starts with happy mood. When the first ray of sunshine pours down in the morning, we feel pleasant temperature and soft light and thus sense of comfort begins. The design has made the most perfect interpretation of this concept, which is to pursue genial sunlight to the fullest.

At the entrance is a long passage rich in depth. No matter the living room, or the bedroom, or the study room, the overall impression the house makes on us is bright and clear window and table, free of dust. Good natural lighting condition, together with pure colors like brown, black and ash, presents a simple and spacious living space.

The use of polished quartz, tawny mirror and black mirror makes light refract and reflect, which have visually enlarged the space and enhanced transparency. Floor of the study room is separated from the living room with a wall, showing the design has abandoned the see-through-at-a-glance layout. You can bathe the warm sunshine on a leisure afternoon, hold a book, drink a cup of tea and find your memories of old days in the fragrance of book and tea. It is sure for you to pass a nice afternoon. The table and chairs in the dining room are seemingly simple and rough but they can bring people illusions of returning to the nature, longing for a feast. Details of the interior design are also exquisite and delicate. Dark red pendulum of the floor clock and rose colored flowers have added into the quiet space a pretty touch of colors. Let' s enjoy the beautiful life quietly!

美好的一天源于愉悦的心情，当清晨第一缕阳光倾泻而下的时候，充分感受到宜人的温度和柔和的光线，舒适感油然而生。这个设计就给了这个理念最完美的诠释——最大限度地追求了和煦的日光照射。

进门处是一条悠长的通道，富有层次感。无论是客厅、卧室还是书房，整体给人的感觉是窗明几净、纤尘不染。良好的采光配合着棕、黑、灰粉等干净的色彩，营造出简约开阔的居家环境。

抛光石英石、茶镜、黑镜的运用，使光线折射反射，扩大了空间，增加了透明度。书房地台与客厅仅一墙之隔，摒弃了一览无余的格局。可以在某一个闲暇的午后，沐浴着暖暖的阳光，捧一本书，品一杯茶，在书香、茶香中寻找尘封的记忆，美好的感觉不言而喻。餐厅内看似简单、粗糙的桌椅却能带给人们回归自然，有种渴望饕餮盛宴的错觉。室内的细节设计精致细腻，暗红色的坐地钟摆、玫瑰色的花团作为点缀都给这恬静的空间增添了一抹俏丽的色彩。让我们静静地享受生活的美好吧！

Miao's Residence

苗公馆

设计师：苗贤哲　设计单位：竹贤设计　项目地点：台湾

This case applies dark grey as its main tone, in order to avoid excessive gloominess , designer adds many of bright elements into interior design, which make it look chic and elegant. Details are emphasized and well organized as well, such as a bouquet of flowers, a candle and a mirror etc. all creating a cozy and romantic atmosphere. The space mixes the senses of masculine and warmth together, in which the house owner could relax and enjoy an exquisite taste.

本案例采用深灰色为主色调，为了避免过于暗淡，设计师又加入了许多明亮的元素，在调节色彩的同时，也让居室更显别致与典雅。设计注重突出细节的构造，一束鲜花、一盏烛台、一面方镜都安排得恰到好处，营造出温馨、浪漫的空间氛围，既有着刚强的理性元素，又散发着细腻甜美的温暖气息。主人回家后，可以得到全身心放松，品味独一无二的生活。

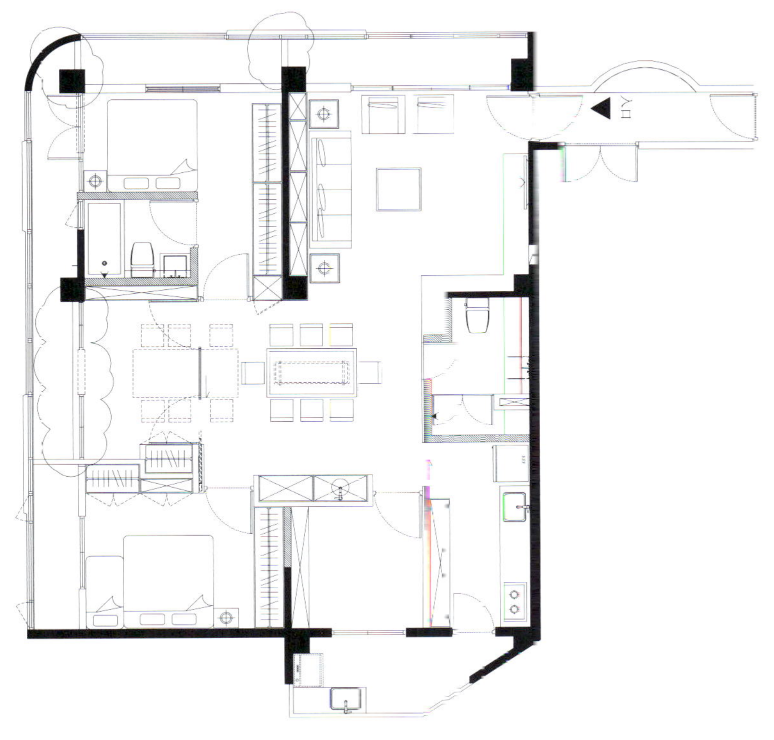

Grass Hill Water Beauty

草山水美

设计师：江姿莹、黄严仕 设计单位：颉合设计

The spatial fluidity makes it possible to get through and combine the indoor and outdoor space. Matching with the mountain views outside and light conditions, the vertical and horizontal axles and layers are created and put in the areas of long tables and sofa by closing or opening the hidden sliding doors at two sides of entrance. The tall book shelf is near the long table, and the specially designed drawers are just fit for the scripts. Here the younger brothers can often discuss plots with directors. Big French window frames the Beitou Da Tun Mountains reflected by the pool in the balcony. The background wall of sofa is made up of DVD cabinets, allowing brothers to enjoy the fun of collecting. When friends come, the solid wooden patterned balcony connects two ends and forms a complete party space.

Natural spring areas of Beitou, the combination of earth, wood and water, natural sculptured stones, solid wooden screen, and shower from the sky can let owners calm down and relax after communicating with nature.

Space is full of memories. It presents the habit of users. Through the details of objects in life, the whole space tells the perfection and characters of materials themselves and satisfies the owners from the aspect of needs and tastes.

空间的流动性使贯穿、融合建筑室内与户外成为可能。在布局上，顺应窗外山景与采光条件，拉出垂直和水平的轴线与层次，置入长桌区及沙发区，利用入口玄关两侧隐藏式的滑门连接或关闭。长桌区旁的全高书柜，定制的抽屉是剧本专用，以便不定时地和导演、编剧讨论剧情；大面的落地窗，将北投大屯山风景引入室内。沙发区背墙是整面的DVD柜，让喜爱电影的兄弟两人可尽情收藏。朋友们来时，柚木实木拼贴的长廊露台串联起两端，成为完整的宽敞的聚会空间。

拥有北投天然温泉的汤区，利用土、木、水的结合，天然凿痕的石材表面，柚木屏风，从天而降的雨水，充分沉淀心灵。经过与自然的对话，彻底放松。

空间是充满记忆的，它清楚地体现居住者的使用习惯。整个空间通过对象细节的呈现，诉说材质本身的美与特性，并满足屋主精神上的需求与品位的彰显。

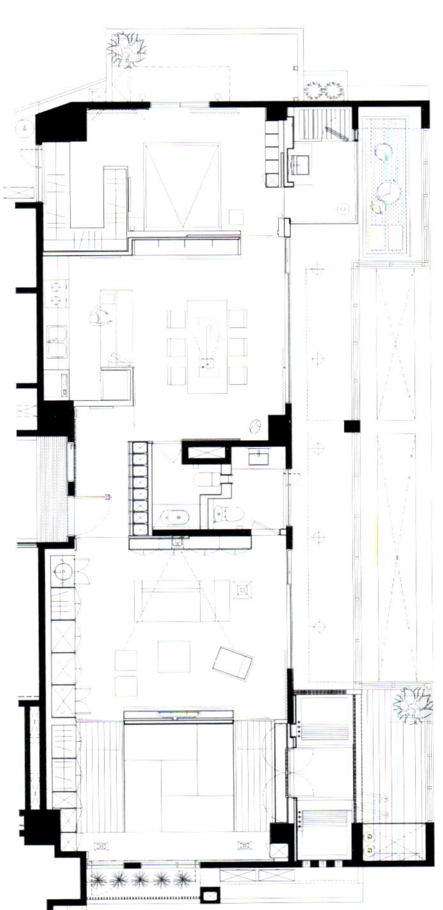

Life Full of Oxygen Like in Forest

涵氧森活

执行设计师：黄鹏霖（Janus Huang）、黄怀德（Roy Huang） 参与设计师：黄恺钧（Dash Huang） 设计单位：TBDC台北基础设计中心 项目地点：台湾台北 建筑面积：约66平方米 主要材料：环保地板、梧桐木、砖、防水涂料 摄影师：王基守

People lead a busy life in the city. Besides basic life functions, the pressure relieving function of a living space are attracting more and more attention. This project is located in the neighborhood of the Bailusi Mountain. So how to bring the environment advantage naturally into the interior space is an important issue to consider in the design. At the initial stage of design, the designers have given up thinking about shapes. Through analysis of the environment advantage, designers have decided that day lighting, ventilation, wood material, stone and sound of the Bailusi Mountain should be the main constructing elements of the space.

The designers made use of the large wardrobe as the partition between living room and master bedroom. One side of the partition is made of natural wood material to stay in harmony with exterior environment; the other side of it is a large area of vacant surface to set a stage for natural light to shine upon. Partition of the bathroom is made of brick wall, whose rough texture is reserved. Flexible partition and sliding door are also treated with natural wood material to make the sounds in the natural environment free to enter. Environmentally friendly non-toxic floor is used to make the users feel like being in the nature although in the interior space.

The space is a relatively small one. To meet the demands of the users, there should be a living room, a dining room, two bedrooms, a kitchen, a bathroom with wet and dry separation, a laundry room, a working balcony and a study room. When dividing the functional spaces, the designers hoped to maintain ventilation to the fullest extent.

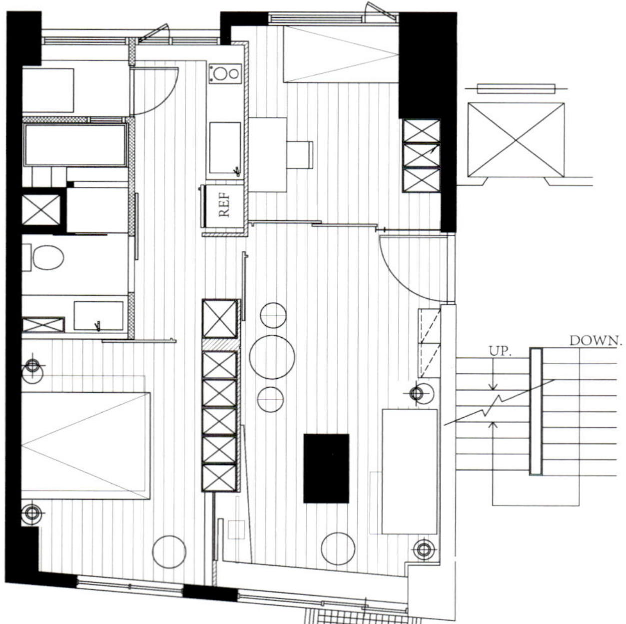

Although functions are clearly defined, spaces reserved inside the wall and the use of large sliding door and flexible partition make it possible that there is no obstruction for light and sound to pass and flow freely in the limited space, ensuring that users have the greatest freedom when living in the space. As to the plan of lighting, apart from plenty of natural light, energy-efficient LED lighting is used at night. Other functional lighting has adopted movable lights.

Living in such a space, one can see a green sea of trees, hear birds and bugs singing and feel the gentle touch of breeze, instead of noisy city streets. One can enjoy calm and pleasant breath and the sounds on the Bailusi Mountain.

都会生活忙碌，居家空间除了满足基本生活的功能外，其纾压功能也越来越受到重视。此案例紧邻绿意盎然的白鹭鸶山，如何将环境条件的优势，自然地延续至室内空间，成为此案例设计的思考重点。设计初期，设计者放弃造型的思考，通过对环境优势的分析，利用采光、空气流动、木料、石块及来自白鹭鸶山的声音，作为空间的主要构成要素。

设计中将大衣柜作为客厅及主卧的隔断，一面使用天然木料，与室外环境呼应，另一面大面积留白，为自然光提供挥洒的舞台。浴厕空间隔间使用砖墙隔开，表面保留斑驳肌理。自主收放的活动隔间及拉门，也使用天然木料处理，让自然环境的声音不受限制。地面使用环保、无毒的地板，尽管是在室内，也让使用者与大自然亲密接触。

虽面积有限，但应使用者要求还要容纳客厅、餐厅、两间卧室、厨房、干湿分离的浴厕空间、洗衣房、工作阳台，以及书房等功能。在空间功能分割的部分，设计者希望保留空间最大的穿透性，功能虽被明确界定，在墙体预留出空间，以及大面移门与活动隔间的运用，让光线引导视觉，甚至是声音的传递，都能在有限的空间内不受局限，自由地穿越与流动，确保使用者在使用的过程中，获得最大的自由感。照明计划部分，除大量引入自然采光外，夜间时除局部情境需要使用节能LED照明，其余功能性照明则利用活动灯具。

树海绿荫，鸟叫虫鸣，微风轻拂，取代喧闹的都会丛林，平静愉悦的呼吸声与白鹭鸶山的声音，谱出新的人生乐章。

Necessary Knowledge of Small Space Design
空间实学

执行设计师：黄鹏霖（Janus Huang）、黄怀德（Roy Huang） 参与设计师：黄恺钧（Dash Huang） 设计单位：TBDC台北基础设计中心 建筑面积：约59.5平方米 项目地点：台湾台北 主要材料：防滑石英砖、环保地板、不锈钢、夹膜玻璃、天然大理石、橡木钢刷木皮染色、橡木实木 摄影师：王基守

Restrictions are necessary conditions for design. Starting from the users' functional needs and making the limited space flexible are designers' main concerns at the initial stage.

At the beginning of the design, through several times of rational and in-depth communication, designers have comprehended the users' routine of the day, storage items, audio-visual needs, aesthetic judgments, nature of work, personality and preferences and so on. Apart from that, designers have also investigated conditions of space area, ventilation, lighting, height limitation and so on and made a configuration consisting of the entrance room, living room, kitchen, master bedroom, bathroom, vertical dynamic line and work space at the back balcony. Because the host and the hostess need independent working spaces, the designers have made a layout plan that contains separate study room, work studio and guest room, which maintain clear division and independent functions. At the same time, making the functional spaces overlap form a large open space. With the arrangement of the position of the stair, it can serve as a dynamic partition as entering the master bedroom, bathroom, study room and work studio, making movements in the space smooth and free.

As to the plan of storage and embedding, air conditioner of the living room is placed at the top of the entrance room, the one of the master bedroom is set at the top of the wardrobe, and that of the study room is embedded in the wall. Wood grill cabinets are built in partial areas of the ceiling and the wall, besides the hidden air conditioners and combination of lighting plan, to maximize the ceiling height.The rest of the lighting

is placed in the vertical surfaces or the floor, making various lighting meet the needs of different situations. Lighting of work space has adopted a considerable number of moveable lights. The lowest part of the ceiling is decorated with mirrors to make use of the limited structure to fix the lighting. Glass partition is used in the bathroom to minimize the sense of repression in the space. The original steel structure is exposed in partial area to meet the most basic aesthetic needs.

The shoe case at the entrance is hidden design and is embedded in the wall behind the full-length mirror, to enlarge the space at the entrance for turning around. What' s more, the use of mirror has enabled the vision to extend and reflex.

Place for household electrical appliances has given up traditional TV wall and is approached with metal components design. Audio-visual equipment is embedded with metal box and placed on the platform that function as a side table, to reduce the visual masking.

Vertical moving line has made use of plates of different sizes and serves at the same time as stair, dining table and desk. It also maintains the functions of storage space for household electrical appliances and computer in the work studio. When meeting different needs toward functions, the design has made the space flexible.

The design has raised basic life functions to a new level of pursuit of life quality. To create an aesthetic space is only designer' s basic capacity, but how to do a practical thinking and planning in a limited space is the necessary knowledge when designing a small space.

限制，是设计必要的条件。从使用者的功能需求出发，让有限的空间条件演化出空间的弹性，是设计者在初期思考的重点。

在设计初期，通过多次理性且深入的沟通，了解使用者的生活作息、收纳品种、影音需求、美学判断、工作性质及性格喜好等，同时调查空间的面积、气流采光、楼高限制等条件，切割出玄关、客厅、厨房、主卧室、浴室、垂直动线以及后阳台的工作区。因男女主人需要独立空间工作之故，分别切割出书房、工作室兼客房等的分区计划，分区明确且功能独立。同时让空间使用重叠，形成一个大的开放空间，借由楼梯位置的安排，作为进入主卧、浴室、书房及工作室的动线分流，让移动的过程流畅自由。

在收纳与嵌入的规划方面，客厅的空调设置在玄关上方，主卧室的空调设置在衣柜顶部，书房的则设置在墙面内。局部施做天花板及墙面实木格栅的柜体，除了隐蔽空调设备，同时结合照明的计划，争取天花板高度。其余的照明则设置在立面或者地面，让不同的光源满足情境需求，工作照明则大量采用活动灯饰。在天花板最低处，采用镜面的处理，利用有限的结构设置照明，浴室隔间使用玻璃分隔，最大限度地减少压迫感，局部外露原始钢骨结构，满足最基本的美学需求。

玄关鞋柜采用隐藏室的设计，将鞋柜纳入墙面穿衣镜，扩大入口的回转空间，同时通过镜面的使用，让视觉视线得以反射延伸。

客厅家电位置的部分，放弃传统电视墙面，改用金属构件的方式处理，视听设备则设计金属盒嵌入，放置在具备边几功能的平台上，减少视觉的遮蔽。

垂直动线利用不同大小的板块设计，在扮演楼梯、餐桌、书桌的角色的同时具备家电设备收纳，以及工作室计算机设备的收纳整合功能，在满足不同的功能的同时，衍生出空间的变化性。

将基本的生活功能，升华至另一种对生活质量的追求，美学的营造只是设计者本身基本具备的能力，如何在有限的空间中，实事求是地计划思考，才是小空间不可或缺的实学。

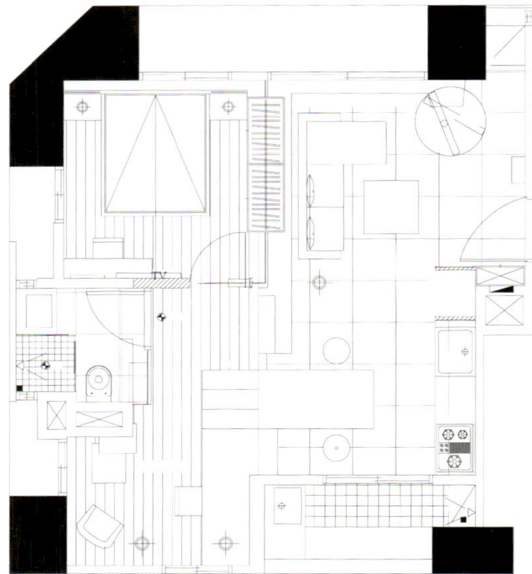

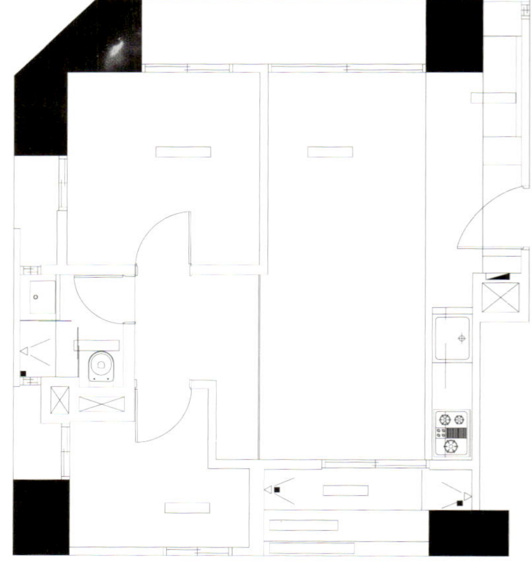

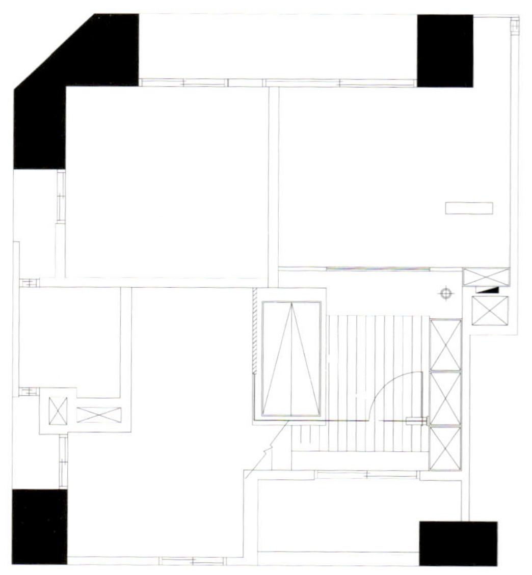

Guo's Residence at East Zhongxiao Road

忠孝东路郭公馆

设计师：周建志　设计单位：春雨时尚空间设计　项目地点：台湾台北　建筑面积：92平方米

The magic world is coming into view at the moment you open the door. The shoe cabinet is used to widen the space here. For example, the big top arch shoe cabinet can store most collections. Meanwhile it is used as the screen separating the passage at the back. There is a large storage room at the back of big tawny mirror on the left side. The walls with little cobbles and the ancient cooper bricks on the floor add much holiday atmosphere.

Compared with the thick lines of furniture, the wall papers with changeable and soft lines are the key elements for the spatial expression. The curved aisle is designed to cover the pipes and pillars and to extend the space visually. The light color wall papers along the narrow passage add much dynamism and widen the narrow space. Along the curved passage, the wall papers are still used to decorate the dining room. The bright color wall papers extend from the walls to the ceiling in L shape and the arch curves make the space higher and wider visually.

设计师的空间魔法在推开大门那一刻，就以造型鞋柜加大空间感的手法呈现，至顶的圆弧大鞋柜完整收纳屋主的大量收藏，也成为与后方廊道隔开的屏风，拉开左侧的大片茶镜，后方是超大储藏空间，细石子的壁面与地面复古锈铜砖丰富度假气氛。

较之厚实的家具线条，多变且线条柔和的壁纸成为铺陈空间表情的主角。为了包覆管道及柱体，设计师以曲折的走道设计延长视觉感受，窄浅的走道空间以浅色调壁纸铺面，增加空间动感，加大廊道空间。顺着曲折廊道入内，设计师继续以壁纸铺陈餐厅的空间表情。对比色度较高的壁纸以L形从壁面延伸至天花空间，圆弧的曲线让空间更高更宽敞。

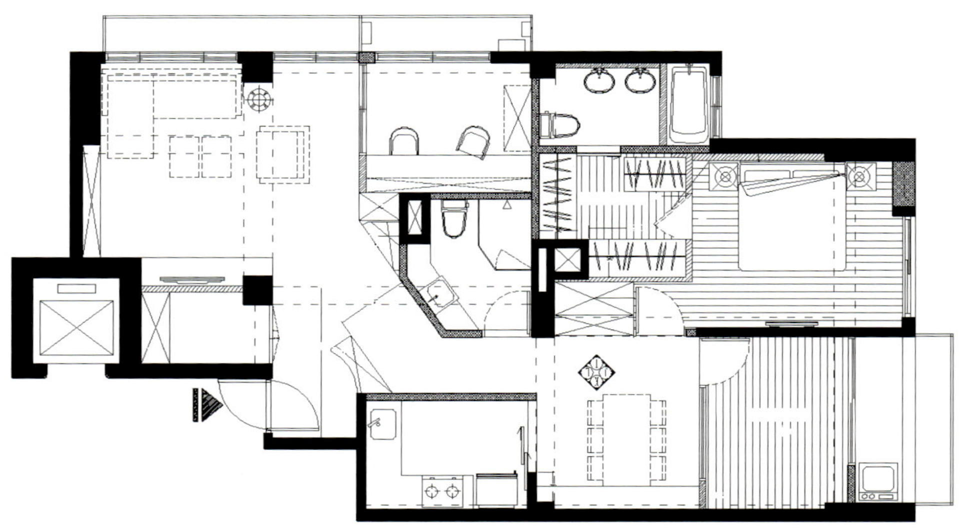

Peaceful Century C3-5F

和平世纪 C3-5F

设计师：江欣宜、吴信池　参与设计：黄建宪、张嘉镁　设计单位：缤纷设计 L' atelier Fantasia　项目地点：台湾新北　建筑面积：38平方米　主要材料：透光人造石、镜面不锈钢、进口壁纸、造型古典线板、钢刷木皮

Black and white are the main colors. The steady black and pure white colors are used creatively. The grey sofa, black and white pillows, and white carpet with black stripes are harmonious. The transparent and artificial stone adopted to the background of TV wall is crystal. The cabinets on both sides can not only display the beloved collections, but also enrich the visual expressions. The black chairs and white round table in the dining room are steady and grand. The decorative lights in the dining room are very special, connected in the form of bubbles stretching from the ceiling to the table. The roman pillars made of transparent artificial stone in the passage and the doors in the bedrooms in the form of lattice extend the space visually, which gives an illusion being in the heaven.

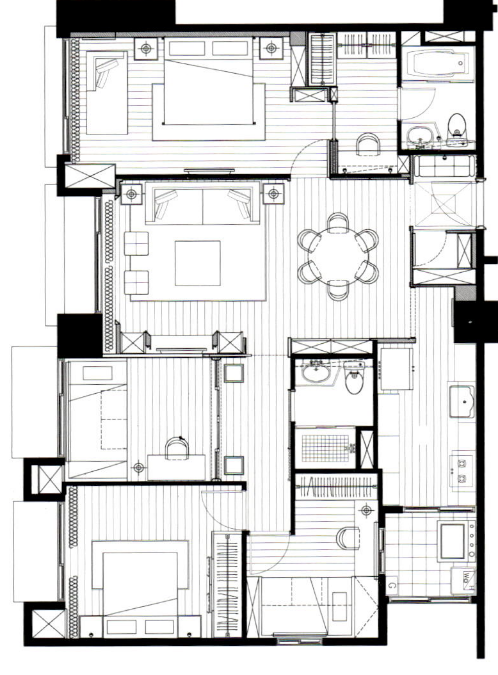

此案例以黑色和白色为主调。设计师用稳重的黑和纯净的白来演绎现代的家居生活，富有创意大胆、追求前卫的个性。灰色的布艺沙发，黑色、白色抱枕，黑条纹的白色地毯与整个色调和谐。设计师以透光人造石打造电视背景墙，晶莹剔透，两边的柜子既可以摆放主人的心爱收藏，也可丰富视觉空间。餐厅黑色餐椅配白色圆桌，沉稳大气。餐厅的装饰性灯具极具个性，以气泡的形式串联，从天花延续到餐桌，给人梦幻迷离的感觉。

走道的透光人造石材质的罗马柱、格子形式的卧室门，都让空间在视觉上得到延伸，给人步入梦幻天堂的错觉。

Peaceful Century D5-5F

和平世纪 D5－5F

设计师：江欣宜、吴信池　参与设计：黄建宪、张嘉镁　设计单位：缤纷设计 L'atelier Fantasia　项目地点：台湾新北　建筑面积：约63平方米　主要材料：意大利弗洛钢石、钢刷栓木木皮、茶镜、不锈钢梯脚、茶玻、进口绷布、定制铁件　摄影师：KYLE、SAM

You can feel the magnificence at the moment when you enter it. The big full-length tawny mirror can not only be used as dressing mirror, but also extend the space. On the sofa wall a black-and-white painting with a wide road spreading far away extends the room visually and also triggers imagination. The ceiling of the dining room is decorated with crystal bottles through which the light is reflected to every corner. The wall of passage is also decorated with tawny mirror extending the space. The background wall in the main bedroom is made of the white squares just like pieces of interesting jigsaw puzzles. The light and bright white yarn curtains create a transparent and bright room.

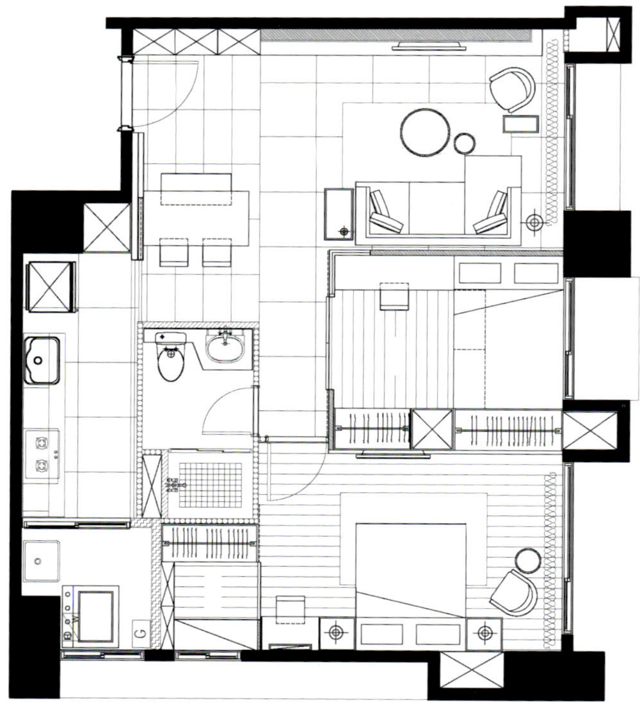

一进玄关门就能感受到空间的大气，整面的大型落地茶镜，除了可作为主人出门时的穿衣镜，也有放大、延伸空间的效果。在沙发背景墙上，一幅一人在宽广的道路向远处走去的黑白画作，不仅在视觉上延伸了空间，也引发了人们的遐想。餐厅的天花上以水晶瓶予以装饰，晶莹剔透的水晶瓶将灯光折射到各个角落。走道的墙面也以茶镜装饰，延伸了空间。主卧的背景墙以白色方块拼接而成，就像一块块拼图，趣味十足。白色纱质窗帘，轻盈剔透，使室内更为通透开阔。

PAUL

Southport Century C1-8F

南港世纪汇C1-8F

设计师：江欣宜、吴信池　设计单位：缤纷设计 L'atelier Fantasia　项目地点：台湾新北　建筑面积：约208平方米　主要材料：客制大理石拼花、木皮、木地板、古典线板、灰镜、茶镜、铁件、贝壳板、进口壁纸、进口家具、手工画框、全室自动化控制系统　摄影师：KYLE、SAM

It is a modern house with comfort and warmness created in a neoclassical way. The grey as main color of the space is idle and significant. The TV wall is made in the form of European fireplace at two sides of which the golden decorations are low-key and luxurious. The drawing room connects the study room in which the bookcase features a sliding mirror door. The door extends the space when it is closed; when opened, the many books inside tell us the owner is well-cultivated. The green paintings on the wall of dining room echo with the potted plants inside, which makes you feel like having dinner in the wonderful spring world. The European droplights and black leather tables create a European and romantic house and enhance the taste of the owner. You will enjoy the quiet and pleasant life when entering.

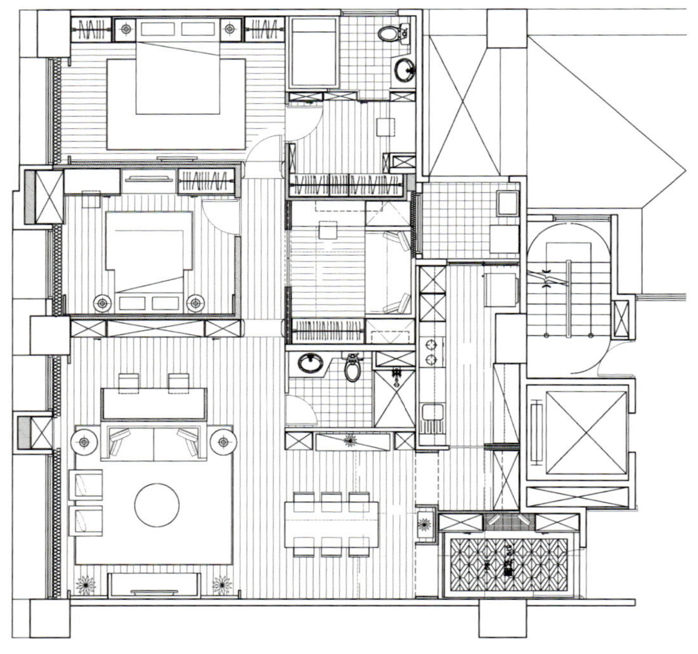

本案以新古典主义的手法打造了一个温馨舒适的现代居所。整个空间以灰色为主色调，透露出慵懒、大气的气质。电视背景以古欧的壁炉为造型，两边造型奇特的金色装饰物赋予空间低调奢华的气息。客厅与书房相连，书架的滑动镜门，关上是一面大镜子，延伸了空间；打开，才发现内有乾坤，满满的藏书显示了主人的文化修养。餐厅墙上的绿色画作与室内的盆栽相映成趣，好似在春天的世界中用餐，心情也得到愉悦。古欧的吊灯、黑色皮质餐椅，浓浓的古欧气息弥漫整个空间，也提升了主人的生活品位。从玄关一踏入此情境，就开始享受安静自在的都会生活。

Lin's Residence, Da Zhi

大直林宅

设计师：张书源　设计单位：丰彤设计　项目地点：台湾台北　建筑面积：室内148.5平方米，户外49.5平方米　主要材料：紫檀地板、南方松、大理石、板岩

Chang Shuyuan who is the designer of this case from Fontal Space Design, re-organized the layout of this irregularly shaped space ingeniously. The original kitchen is changed into a porch, and divides the whole space into public area and private area. Replacing the door, the wall built of glass blocks leads sunshine into elevator and porch of apartment. Due to this change, the gloomy entrance becomes brighter than before. The chic porch creates a great vibe for this irregularly shaped space.

设计师大胆地化整为零，更改入门动线，原厨房位置改为全新玄关，并以此作为分界线，将室内空间界定成公、私领域两大区块，使得彼此得以完整、独立，不互相影响。于是原先的大门位置改成由玻璃砖砌成的墙面，引导屋外阳台的光线进入公寓的电梯间及玄关，使得原本阴暗的入门处变得明亮。特意设计的玄关使原本奇零的空间显得大气。

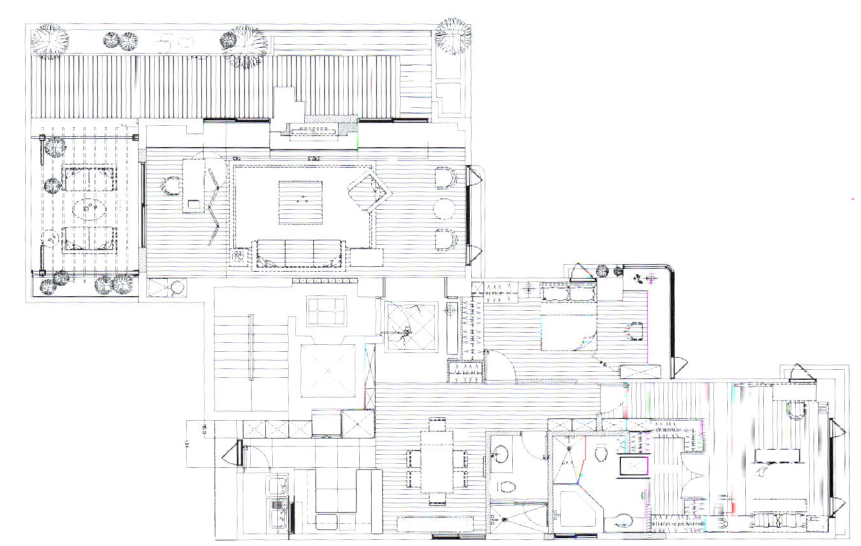

The Hanhuang Masterwork

汉皇极品

设计师：王文亚　参与设计师：吴惠晴、廖佳仪　设计单位：异国设计　项目地点：台湾台北　建筑面积：320平方米　主要材料：石材、深色玻璃、金属图腾

The case presents a distinguished momentum by the unique exquisite technique. For the appearance of the reception hall, it responds to the building's vocabulary to create a stylish, classical, and luxurious style and a noble bearing with stone, dark glass, and metal totems.

This case takes the fashionable and luxurious style and avoids the traditional ornate and magnificent palace-style. The designers use the main tone of black and white, dark stone, leather, dark glass and totems to create a gorgeous sense with individuality. With the advantage of a 3.6 m height, the use of the space is relatively abundant.

The living room, dining room, and the study adopt the approach of sharing space and flexible compartment, which creates a large-scale and generous spatial sense. The open refreshment area and the closed stir-frying area are linked together. And the large multi-purpose bathroom is equipped with Spa equipment and the lower plate bathtub. The separation of wet and dry space indicates its momentum of a luxurious residence.

为突显其尊贵的气势，用独特精致的手法包装该案。接待会馆的外观造型，呼应该建筑物的语汇，打造时尚、古典、奢华的风格，用石材、深色玻璃、金属图腾营造尊贵的气度。

此案例采用时尚、奢华的风格，避免流于传统宫廷式的华丽与金碧辉煌。设计师运用黑白的主调，深色石材与皮件，深色玻璃及图腾，去营造一种有个性的华丽感。该案例有3.6米高度的优势，故在空间的利用上也比较丰富。

在客厅、餐厅、书房采用空间共享和弹性隔间的手法，营造大尺寸、大气度的空间感；开放式的轻食区与可封闭的热炒区连接起来；大尺寸、多功能的浴室，有温泉机设备，配合降板浴缸和干湿分离的空间，突显豪宅气势。

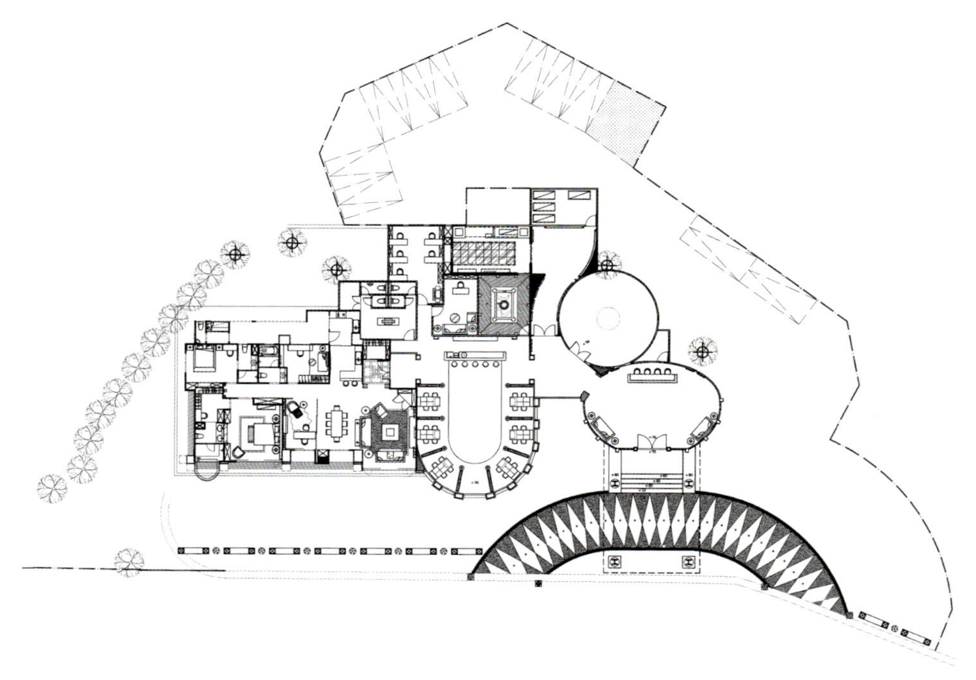

图书在版编目（CIP）数据

港台当红设计师样板房II / 徐宾宾 主编. —南京：
江苏人民出版社，2012.2
ISBN 978-7-214-07652-6

Ⅰ. ①港… Ⅱ. ①徐… Ⅲ. ①住宅—室内装饰设计
—香港—图集 ②住宅-室内装饰设计-台湾省-图集 Ⅳ. ①TU241-64

中国版本图书馆CIP数据核字(2011)第240589号

港台当红设计师样板房II 徐宾宾 主编

责任编辑 蒋卫国 赵 萌
责任监印 彭李君
版式设计 吴华
出　　版 江苏人民出版社（南京湖南路1号A楼 邮编：210009）
发　　行 天津凤凰空间文化传媒有限公司
销售电话 022-87893668
网　　址 http://www.ifengspace.cn
集团地址 凤凰出版传媒集团（南京湖南路1号A楼　邮编：210009）
经　　销 全国新华书店
印　　刷 利丰雅高印刷（深圳）有限公司
开　　本 965毫米×1270毫米　1/16
印　　张 21
字　　数 168千字
版　　次 2012年2月第1版
印　　次 2012年2月第1次印刷
书　　号 ISBN 978-7-214-07652-6
定　　价 298.00(USD 52.00)